Lecture Notes in Mathematics 1674

Springer
Berlin
Heidelberg
New York
Barcelona
Budapest
Hong Kong
London
Milan
Paris
Santa Clara
Singapore
Tokyo

G. Klaas C. R. Leedham-Green
W. Plesken

Linear Pro-p-Groups of Finite Width

Springer

Authors

Gundel Klaas
Wilhelm Plesken
Lehrstuhl B für Mathematik
Templergraben 64
D-52062 Aachen, Germany
e-mail: gz@momo.math.rwth-aachen.de
plesken@momo.math.rwth-aachen.de

Charles R. Leedham-Green
Queen Mary and Westfield College
University of London
School of Mathematical Sciences
Mile End Road
London E1 4NS, England
e-mail: C.R.Leedham-Green@qmw.ac.uk

Cataloging-in-Publication Data applied for

Die Deutsche Bibliothek - CIP-Einheitsaufnahme

Leedham-Green, Charles Richard:
Linear pro-p-groups of finite width / C. R. Leedham-Green ; W. Plesken ; G. Klaas. - Berlin ; Heidelberg ; New York ; Barcelona ; Budapest ; Hong Kong ; London ; Milan ; Paris ; Santa Clara ; Singapore ; Tokyo : Springer, 1997
(Lecture notes in mathematics ; 1674)
ISBN 3-540-63643-9

Mathematics Subject Classification (1991): 20G25, 20E18, 11E95, 20D15

ISSN 0075-8434
ISBN 3-540-63643-9 Springer-Verlag Berlin Heidelberg New York

Typesetting: Camera-ready TEX output by the authors
SPIN: 10553380 46/3142-543210 - Printed on acid-free paper

Preface

The present notes are the result of a research project supported by the DFG (Schwerpunkt: "Algorithmische Zahlentheorie und Algebra"), which has been going on for several years. In previous projects on classifications of p-groups according to coclass a complete structure theory has been developed. The study of pro-p-groups of finite width in these notes is a natural continuation of the coclass project. It revealed many new examples of pro-p-groups. The rôle of p-adic space groups has been replaced by open pro-p-subgroups of semisimple algebraic groups over the p-adic numbers.

While preparing these notes we got help from many people, in particular program support from Colin Murgatroyd and Matthias Zumbroich. Lots of ideas were discussed with B. Souvignier, G. Nebe and R. Camina. The unknown referee, S. Sidki, and R. I. Grigorchuk helped us with valuable references and comments. We thank them all.

Aachen, June 1996

C. R. Leedham-Green, W. Plesken, G. Klaas

Contents

I Introduction

a) Width and just infinite pro-p-groups

Let p be a prime. For a finite p-group P with lower central series

$$\gamma_1(P) := P \geq \gamma_2(P) := [P,P] \geq \cdots \geq \gamma_i(P) := [P, \gamma_{i-1}(P)] \geq \cdots$$

define the *width* $w(P)$ of P by

$$w(P) := \max_i \log_p(|\gamma_i(P)/\gamma_{i+1}(P)|).$$

To study all finite p-groups whose width is bounded by some number b is a rather ambitious project. Here we attempt to approach the easier task of classifying certain infinite pro-p-groups $P := \varprojlim P_i$ whose width $w(P) := \lim w(P_i)$ is bounded by b say. Though P is the inverse limit of a sequence of finite p-groups P_i, the idea of our approach is not to use this as a construction of P, but to construct and investigate P by utilising factor groups of P. As we shall soon see, classifying in this context can no longer mean enumerating the isomorphism types and developing recognition mechanisms, but rather supplying a sufficiently accurate description for one to be able to answer reasonable questions about the structure of these groups. Rather than with the width, we shall sometimes work with the average width, ultimate width or upper average width.

(I.1) Definition. *Let P be an infinite pro-p-group.*

(i) P is said to be of finite width if the width

$$w(P) := \sup_i \log_p(|\gamma_i(P)/\gamma_{i+1}(P)|)$$

is finite.

(ii) If P is of finite width, its average width $w_a(P)$ is defined as

$$w_a(P) := \lim_{i\to\infty} \frac{\log_p(|P/\gamma_{i+1}(P)|)}{i}$$

if the limit exists. If P is not of finite width or if $\log_p(|P/\gamma_{i+1}(P)|)/i$ converges to infinity, then $w_a(P) := \infty$.

(iii) The ultimate width $\overline{w}(P)$ of P is defined as

$$\overline{w}(P) := \overline{\lim_{i\to\infty}} \log_p(|\gamma_i(P)/\gamma_{i+1}(P)|).$$

(iv) The upper average width $\overline{w}_a(P)$ is defined as

$$\overline{w}_a(P) := \overline{\lim_{i\to\infty}} \frac{\log_p(|P/\gamma_{i+1}(P)|)}{i}.$$

Note that all these variants of the width are infinite if P is not finitely generated as a pro-p-group. Whenever the average width $w_a(P)$ is defined for a pro-p-group P, it is equal to the upper average width $\overline{w}_a(P)$. According to [Roz 96] the pro-2-completion P of the Grigorchuk group, cf. [Gri 80], satifies

$$log_2(|\gamma_n(P)/\gamma_{n+1}(P)|) = \begin{cases} 3 & \text{if } n = 1 \\ 2 & \text{if } 2^m + 1 \leq n < 3 \cdot 2^{m-1} + 1 \\ 1 & \text{if } 3 \cdot 2^{m-1} + 1 \leq n < 2^{m+1} + 1 \end{cases} .$$

So it has width 3, upper average width 5/3, and its average width is undefined.

Note that the infinite pro-p-groups P of finite coclass are by definition the groups of ultimate width $\overline{w}(P)$ equal to 1. Obviously their average width is also 1. Pro-p-groups of finite coclass are known to be soluble, cf. [LeN 80], [Don 87], [Sha 94]. At the present time it is not known whether the pro-p-groups of average width 1 are of finite coclass.

(I.2) Lemma. *If P is a pro-p-group, and N is a normal subgroup of P, then $w(P) \geq w(P/N)$, $w_a(P) \geq w_a(P/N)$, $\overline{w}(P) \geq \overline{w}(P/N)$ and $\overline{w_a}(P) \geq \overline{w_a}(P/N)$, whenever these invariants are defined.*

This result is entirely trivial. Note that $w(P)$ is always defined, and $\overline{w}(P)$ and $\overline{w}_a(P)$ are defined if and only if P is infinite.

As in the case of pro-p-groups of finite coclass it is natural to restrict oneself to the just infinite groups.

(I.3) Definition. *An infinite profinite group P is said to be just infinite, if P has no non-trivial (closed) normal subgroup of infinite index, i.e. P has no proper infinite (topological) factor groups.*

It is easy to see, cf. Proposition (II.5), that any finitely generated infinite pro-p-group has a just infinite homomorphic image.

The just infinite pro-p-groups play a similar rôle in the theory of pro-p-groups that is played by simple groups in the theory of finite groups. The similarity is quite marked. In both cases, most examples are of Lie type, with interesting exceptions.

The soluble just infinite pro-p-groups are easily seen to be irreducible p-adic space groups. That is to say, groups P with a normal subgroup T, where T is a finitely generated free $\mathbb{Z}_p$-module, and P/T is a finite p-group that acts faithfully and irreducibly on T. To investigate them some of the machinery developed to study pro-p groups of the finite coclass can be used. Here we concentrate on insoluble just infinite pro-p-groups of finite width. Their average width can come arbitrarily close to 1, which might be unexpected because pro-p-groups of finite coclass are soluble.

(I.4) Theorem. *Let $b > 1$ be a real number. Then there exists an infinite insoluble pro-p-group P of finite average width with $1 < w_a(P) < b$.*

This result is proved in Chapter VII. To prove the theorem with 'insoluble' replaced by 'soluble' is left as an exercise to the reader.

b) Ultimate periodicity and obliquity

The second issue of these notes is the ultimate periodicity of the sequence $(|\gamma_i(P) : \gamma_{i+1}(P)|)_{i\in\mathbb{N}}$ for insoluble just infinite pro-p-groups P of finite width. The only example known up to now not to have this property is the pro-2-completion of the Grigorchuk group, cf. [Roz 96], [Gri 80]. If the property holds with

$$(|\gamma_i(P) : \gamma_{i+1}(P)|) = (|\gamma_{i+t}(P) : \gamma_{i+t+1}(P)|)$$

for $i \geq i_0$ and some $t \in \mathbb{N}$, the average width is defined and equal to the arithmetic mean of $\log_p(|\gamma_i(P) : \gamma_{i+1}(P)|)$ for $i = i_0, \ldots, i_0 + t - 1$. Again proving ultimate periodicity in the case of soluble pro-p-groups of bounded width is an easy exercise. In all cases investigated up to now, ultimate periodicity of the sequence $(|\gamma_i(P) : \gamma_{i+1}(P)|)_{i\in\mathbb{N}}$ is only part of the periodic pattern. One usually has for some $t > 0$ an isomorphism of the lattice of open normal subgroups of P contained in $\gamma_i(P)$ onto the lattice of open normal subgroups of P contained in $\gamma_{i+t}(P)$ mapping $\gamma_j(P)$ onto $\gamma_{j+t}(P)$ for $j \geq i$, cf. Chapter III. Closely related is the question of how restricted the lattice of normal subgroups is. More precisely, we want to measure how far one is removed from the most restrictive situation, where each normal subgroup N of P satisfies $\gamma_{i+1}(P) \leq N \leq \gamma_i(P)$ for some $i = i(N)$. The concept measuring this is called 'obliquity'.

(I.5) Definition. *Let P be a pro-p-group of finite width, and let $i > 0$. Define $\mu_i(P)$ to be the intersection of $\gamma_{i+1}(P)$ with the intersection of all normal subgroups N of P with $N \not\leq \gamma_{i+1}(P)$. The ith obliquity o_i is defined as $o_i(P) := \log_p(|\gamma_{i+1}(P) : \mu_i(P)|)$. Define the obliquity $o(P)$ of P to be $max\{o_i(P) \mid i \in \mathbb{N}\}$ if the maximum exists; otherwise set $o(P) = \infty$.*

It is easy to see that if P is a pro-p-group of finite width with a normal subgroup $N \neq \langle 1\rangle$ of infinite index in P then P is not of finite obliquity. Thus the obliquity of P can be seen as a numerical invariant that gives more precise information than the simple question of whether or not P is just infinite. However, we do not know whether or not every just infinite pro-p-group of finite width has finite obliquity.

It may occur to the reader that it would be more natural to define the width of a pro-p-group in terms of arbitrary central sections, rather than lower central sections. This approach has theoretical advantages, but suffers from the difficulty that the width would then be harder to calculate. We shall see that if P has finite width (in our sense) and finite obliquity, then there is a uniform bound to the orders of the central sections of P, cf. Lemma (II.3). However, if P is a just infinite pro-p-group of finite width, and P is not finitely presented as a pro-p-group, then the p-covering group P is of finite width, but has an infinite centre.

(1.6) Lemma. *Let P be a pro-p-group of finite width, and let N be a closed subgroup of P. Then $o(P) \geq o(P/N)$.*

Proof. This follows from the fact that $\mu_i(P/N) \geq \mu_i(P)N/N$. q.e.d.

Turning to finite p-groups, we have seen that for given p, o and w the p-groups of width $\leq w$ and obliquity $\leq o$ form a quotient-closed class of groups. Let $\Gamma_{p,w,o}$ be the graph whose vertices are the isomorphism classes of the p-groups of width $\leq w$ and obliquity $\leq o$. Confusing a group with its isomorphism class, if $G \in \Gamma_{p,w,o}$ is of class c, join G to $G/\gamma_c(G)$. Then every just infinite pro-p-group of width at most w and obliquity at most o is the inverse limit of an infinite chain in $\Gamma_{p,w,o}$.
Thus we might hope to start a classification of finite p-groups by constructing all just infinite pro-p-groups of finite width, and taking the finite homomorphic images of these groups as our prime source of examples. Unfortunately, as we shall see, there are uncountably many pro-p-groups in this class. However, if we also bound the subgroup rank of the finite groups, we then find that there are only finitely many pro-p-groups to deal with (for fixed p, and with fixed bounds to the width, obliquity and subgroup rank of the finite p-groups in the graph). Moreover, we need to bound all three invariants, width, obliquity and subgroup rank, since if we do not bound the obliquity we will get pro-p-groups that are not just infinite as inverse limits of our finite p-groups, and if we remove the restriction on the width or the subgroup rank, we get infinitely many pro-p-groups. However, we do not wish to exclude groups of infinite subgroup rank completely from our considerations as they are of great interest.

c) Four types of just infinite pro-p-groups

The insoluble pro-p-groups of finite width which are most accessible are p-adic analytic, cf. [LuM 87b], [DdMS 91]. For ease, we propose the following definition.

(I.7) Definition. *A p-adically simple group, for short a $\tilde{p}$-group, is a p-adic analytic just infinite pro-p-group.*

Every $\tilde{p}$-group is of finite width. One expects plenty of other just infinite pro-p-groups of infinite width. Determining the width might be rather difficult, e. g. it seems that the width of the p-completion of the Gupta-Sidki groups (cf. [Sid 84]) are not yet determined. Since these are p-analogs of the Grigorchuk group, the answer would be particularly interesting. In these notes we are mainly concerned with insoluble $\tilde{p}$-groups. One can distinguish four types of just infinite pro-p-groups of finite width: the soluble ones; the insoluble $\tilde{p}$-groups; those that are not p-adic analytic, but are linear over $\mathbb{F}_p((t))$; and those that are non-linear, i.e. the rest, which are not linear over $\mathbb{Q}_p$ or $\mathbb{F}_p((t))$. A comment on each type seems to be appropriate.

The first two types are both $\tilde{p}$-groups, however the difference between the two is quite marked. The soluble ones are irreducible p-adic space groups. Essentially they can be investigated by the methods developed to study pro-p-groups of finite coclass. In fact, Conjecture C of [LeN 80], which has since been proved, cf. [Don 87], [Lee 94a], [Sha 94], says: every pro-p-group of finite coclass is soluble. The insoluble $\tilde{p}$-groups are therefore a new class of groups to be investigated and they form the main topic of this work.

The third type, e.g. Sylow pro-p-subgroups of Chevalley groups over $\mathbb{F}_p((t))$, can often be treated at the same time as insoluble $\tilde{p}$-groups. For example when one computes lower central series. However, this class of groups has many unpleasant properties. They may be isomorphic to proper subgroups of themselves, they may have p-groups of outer automorphisms of arbitrarily big orders, they give rise to uncountably many just infinite pro-p-groups of finite width, cf. Chapter XIII. Also they are not easily investigated by computers, because it is not clear when one has investigated a sufficiently large finite factor group to predict properties of the infinite group e.g. the sections of the lower central series. The attitude taken in these notes is to treat them simultaneously with the insoluble $\tilde{p}$-groups, whenever no extra effort is necessary.

Finally, as for the fourth type, i.e. the non-linear groups, only a few classes of examples are known, namely the Nottingham groups and the p-completion of the Grigorchuk group, cf. [Gri 80], [Roz 96], and their open subgroups. A Nottingham group $S(q)$ is the Sylow pro-p-subgroup of the group of $\mathbb{F}_q$-algebra automorphisms of $\mathbb{F}_q[[t]]$, where q is a power of p, cf. Chapter XIII a). Our perception of the general structure of pro-p-groups of finite width has been radically changed by the recently discovered fact that $S(p)$, for p any prime, contains as a closed subgroup an isomorphic copy of *every* countably based pro-p-group. See [Cam 97].

It seems far more hopeful to try to describe all just infinite pro-p groups P with the property that every open subgroup of P is just infinite. Call these groups 'hereditary just infinite'. The known hereditary just infinite groups of finite width are the just infinite groups described above where the associated Lie algebras (for the second type cf. Chapter III) are simple, and just infinite groups commesurate with the Nottingham group. (Note that G and H are commesurable if they have isomorphic open subgroups.)

d) Non-soluble p-adically simple groups

Our main emphasis is on the groups of type 2; that is, on the insoluble just infinite p-adic analytic pro-p-groups of finite width, or insoluble $\tilde{p}$-groups for short. These occur in families, where a family is an equivalence class under the relation $G \sim H$ if G and H have open isomorphic subgroups. Then it turns out that the relation '$\prec$', defined by $H \prec G \Leftrightarrow H$ is isomorphic to an open subgroup of G defines a partial order on each family, and that every family has a unique maximal group. It is these maximal groups that we exhibit, in various cases.

The reason why insoluble $\tilde{p}$-groups P are accessible in a rather explicit way is that they are open subgroups of the group of $\mathbb{Q}_p$-rational points of certain semisimple algebraic groups G defined over the field $\mathbb{Q}_p$ of p-adic numbers. In the p-adic topology P is necessarily compact and the topology induced from the $\mathbb{Q}_p$-topology of $G(\mathbb{Q}_p)$ is the same as the original profinite topology of P. More precisely, it is possible to attach a $\mathbb{Q}_p$-Lie algebra to P, which is the direct sum of p^α copies of a simple $\mathbb{Q}_p$-Lie algebra for some $\alpha \in \mathbb{Z}_{\geq 0}$ (with $\alpha = 0$ as the most interesting case). The algebraic group G is the automorphism group of this Lie algebra. Therefore the classification of simple $\mathbb{Q}_p$-Lie algebras and simple algebraic groups over $\mathbb{Q}_p$, cf. [Kne 65], [Sat 71], or [BrT 72], [BrT 84], [BrT 87] can be used. The last three references also cover the simple algebraic groups over local fields of positive characteristic which we sometimes also treat, cf. Chapters V and VI. Even if the simple Lie algebra is given, it is not always a trivial exercise to find the maximal $\tilde{p}$-group associated with it, which is a Sylow pro-p-subgroup of $G(\mathbb{Q}_p)$. In practice, the problem of passing from a Sylow pro-p-subgroup of $G^o(\mathbb{Q}_p)$ to one of $G(\mathbb{Q}_p)$ is not always routine. Here G^o denotes the connected 1-component of G in the sense of algebraic groups.

e) Contents and organisation of these notes

Chapters II to VII and Chapter XIII are purely theoretical investigations into just infinite pro-p-groups of finite width. Chapters VIII to XII deal with algorithms, showing how to treat these groups with a computer. Where it was impossible to proceed theoretically, the tables of results of computer calculations have been given in Chapter XII. Finally, Chapter XIV lists some open problems.

Chapter II concerns generalities on pro-p-groups of finite width, which are easy and do not use the distinction between the four types of groups discussed in c) above. Chapter III proves all the general results on $\tilde{p}$-groups indicated in d) above, in particular the interplay between simple $\mathbb{Q}_p$-Lie algebras and insoluble $\tilde{p}$-groups and the existence and uniqueness of maximal $\tilde{p}$-groups. It also discusses the obliquity of $\tilde{p}$-groups. Chapter IV deals with the question of which factor groups of a $\tilde{p}$-group are sufficiently large to determine the pattern of the lower central series. The results are modelled after Blackburn's Theorems for the case $(f_1, f_2, \ldots) = (2, 1, 2, 1, \ldots)$ where $f_i = \log_p(|\gamma_i(P)/\gamma_{i+1}(P)|)$, cf. [Bla 61] reproduced in [Hup 67] p. 392.

The next three Chapters, V to VII, deal with specific groups for which the lower central series is computed. Though the main intention is to deal with maximal $\tilde{p}$-groups, it turns out that the analogues in characteristic p can be treated at the same time. Chapter V deals with the Sylow pro-p-subgroups of (split) Chevalley groups over a local field. For most cases the lower central series is determined. Chapter VI deals with the Sylow pro-p-subgroups of the classical groups over a local field where the characteristic of the residue class field is not 2. It turns out that the Cayley map can be used to compute the lower central series thus avoiding the machinery developed in [BrT 72], [BrT 84], and [BrT 87]. Instead one has to deal with algebras with involution and one can assign a canonical hereditary order to the Sylow pro-p-subgroup, which is invariant under the involution. This order allows one to read off the lower central series. The method fails only for unitary groups of degree 3 over a local field whose residue class field has characteristic 3. As an interesting by-product one gets a formula similar to the Baker-Campbell-Hausdorff formula for classical groups which uses associative polynomials in the algebra with involution rather than Lie polynomials. It might be that this formula can also be used in the case when the characteristic of the residue class field is 2. Finally, Chapter VII deals with the Sylow pro-p-subgroup of $PGL(p^\alpha, \mathbb{Q}_p)$, a case which is not covered by Chapter V on Chevalley groups, since so called rational automorphisms have to be taken into account. In any case, these rational automorphisms lead to groups with a rather small average width. Then Theorem (I.4) follows from the investigations of Chapter VII.

The aim of Chapters VIII to XII is to investigate the lower central series and the obliquity of all insoluble, maximal $\tilde{p}$-groups whose associated Lie algebras have dimension ≤ 14 mainly for primes $p = 2, 3$. It is in these cases that it often becomes difficult to decide whether a given $\tilde{p}$-group is maximal or not, cf. Chapter XI. For $p \geq 5$ the lower central series can be investigated theoretically as described in Chapters V and VI, since up to dimension 14 there are no additional p-automorphisms on top of the Sylow pro-p-subgroups of $G^o(\mathbb{Q}_p)$ with G and G^o as at the end of Section d) above. For $p = 2, 3$ the groups considered in the theoretical chapters are often not maximal $\tilde{p}$-groups. The task of deciding maximality, respectively of finding the maximal $\tilde{p}$-groups, has obviously both a theoretical and a computational aspect. In making these groups explicit, we have faced a number of difficulties.

First, we need to know the local fields of characteristic 0 and their automorphism groups in small cases. Although obtaining this information is a routine computation, we have not been able to find these fields in the literature, and have included them here in Chapter IX.

Secondly, it is not clear how best to make these groups explicit. They are generated, topologically, by a finite number of matrices, together, perhaps, with some Galois automorphisms, etc. These generators are the starting point of our calculations. However, they require a fair amount of space on the page, and are not, in themselves, very informative. We could give a power-commutator presentation for a large quotient of each group. This gives a good insight into the structure of the groups, but is prohibitively expensive in space. A third possibility would be to make generally available the computational tools that we have used. Here we have two problems. First, the code has developed in an *ad hoc* way as the project has advanced. We are very grateful to Colin Murgatroyd and Matthias Zumbroich who wrote a great deal of it. The program, particularly the local field arithmetic, is subtle, complicated and extensive, but it works very efficiently, and appears to be completely reliable. However, it is not entirely portable, and experience has suggested improvements that could be made. We intend to get a publically usable version written; in the meantime, we will be glad to run any examples sent to us. In any case, the routines for doing the field arithmetic fast, which is absolutely necessary even for the examples with a Lie algebra of small dimension, are described in Chapter VIII. Possibly these algorithms are of independent interest. The algorithms which use the field arithmetic (characteristic 0 or p) to compute a power commutator presentation for suitable factor groups of a given group P are described in Chapter X.

Chapter XI gives enough details to obtain generators for the maximal insoluble $\tilde{p}$-groups. Applying the algorithms to a suitable factor group yields a power-commutator presentation of the factor group. Using these presentations for factor groups of a sufficient size (such that the factor specifies the structure of the pro-p-group P) one obtains, by calculations in GAP [GAP 94], the lower central series and the obliquity of P.

In particular, Chapter XII lists the tables of all insoluble maximal $\tilde{p}$-groups for $p = 2, 3$ where the dimension of the associated $\mathbb{Q}_p$-Lie algebra is at most 14. The tables include the pattern of the lower central series and information about the obliquity, i.e. the obliquity or at least the ultimate obliquity. The results show that the task is impossible to master without a computer. For instance the pattern of the lower central series often becomes periodic only after a long preperiod. We hope that these tables not only provide groups that are interesting in their own right but can also be used to investigate general insoluble $\tilde{p}$-groups for small dimensions.

II Elementary properties of width

It is clear that the property of having finite width is closed under the formation of finite direct products, and finite p-extensions. For just infinite insoluble pro-p-groups it is probably also closed under passage to an open subgroup, however has not been proved. The best result we have for passage to open subgroups is the following.

(II.1) Proposition. *Let P be a just infinite, insoluble pro-p-group of finite upper average width and let $Q < P$ be open; then Q is of finite upper average width.*

Proof. Without loss of generality let $|P : Q| = p$. Clearly $\gamma_i(Q) \leq \gamma_i(P)$. Now $\gamma_i(Q)$ is characteristic in Q, and Q is normal in P, so $\gamma_i(Q)$ is normal in P. Therefore $|P : \gamma_i(Q)| < \infty$ or $\gamma_i(Q) = \langle 1 \rangle$. The last possibility is ruled out because Q is insoluble. Hence $\gamma_i(Q)/\gamma_{i+1}(Q)$ is finite for each i. The exponent of $\gamma_i(Q)/\gamma_{i+1}(Q)$ is bounded by the exponent p^α of $Q/\gamma_2(Q)$. Let k be minimal with $A^k = 0$ where A is the augmentation ideal of the group ring $(\mathbb{Z}/p^\alpha\mathbb{Z})C_p$. Then the k-fold commutator group of $\gamma_i(Q)$ with P satisfies $[\gamma_i(Q),_k P] \leq \gamma_{i+1}(Q)$ where, for any group G and $H \leq G$, we define $[H,_1 G] := [H, G]$ and $[H,_{i+1} G] := [[H,_i G], G]$ for $i \geq 1$.
We now prove, by induction on i, that $\gamma_{ki}(P) \leq \gamma_i(Q)$. Clearly $\gamma_k(P) \leq \gamma_2(P) \leq Q = \gamma_1(Q)$ since P/Q is abelian and $k \geq 2$. Assume the claim holds for i. Then $\gamma_{i+1}(Q) \geq [\gamma_i(Q),_k P] \geq [\gamma_{ki}(P),_k P] = \gamma_{k(i+1)}(P)$, which proves the induction hypothesis. So

$$\overline{w}_a(Q) = \overline{\lim_{i\to\infty}} \frac{\log_p(|Q : \gamma_{i+1}(Q)|)}{i} \leq \overline{\lim_{i\to\infty}} \frac{\log_p(|P : \gamma_{k(i+1)}(P)|)}{i} = k\overline{w}_a(P).$$

q.e.d.

(II.2) Lemma. *Let P be an infinite, finitely generated pro-p-group and let Q be an open subgroup of P. Then $\overline{w}_a(Q) \geq \overline{w}_a(P)$.*

Proof. Since $Q \leq P$, one has $\gamma_i(Q) \leq \gamma_i(P)$ and therefore

$$\begin{aligned}\log_p |Q : \gamma_{i+1}(Q)| &= -\log_p |P : Q| + \log_p |P : \gamma_{i+1}(P)| + \log_p |\gamma_{i+1}(P) : \gamma_{i+1}(Q)| \\ &\geq -\log_p |P : Q| + \log_p |P : \gamma_{i+1}(P)|.\end{aligned}$$

Dividing by i and taking upper limits gives the result. q.e.d.

There does not seem to be a version of (II.1) for the width instead of the upper average width. However, the following holds.

(II.3) Lemma. *Let P be an infinite pro-p-group of finite width $w(P)$ and finite obliquity $o(P)$. Then there is a $k \in \mathbb{N}$ such that any central section M/N of P with M and N open normal subgroups of P and $N \leq M$ satisfies $|M : N| \leq p^k$.*

Proof. Let $M, N \trianglelefteq P$ be open with $N \leq M$ and M/N centralised by P. Since $o(P)$ is finite there is a $t \in \mathbb{N}$ independent of M such that the following holds. There exists $i \in \mathbb{N}$ with $\gamma_{i+t}(P) \leq M \leq \gamma_i(P)$. Since M/N is centralised by P, one has $\gamma_{i+t+1}(P) \leq N \leq \gamma_{i+1}(P)$. Hence $|M/N| \leq p^{(t+1)w(P)}$. q.e.d.

(II.4) Corollary. *Let P be an infinite, insoluble pro-p-group of finite width and finite obliquity. Then any open subgroup of P has finite width.*

Proof. By Lemma (II.3) the central sections M/N of P are of order bounded by p^k for some k. Let $Q \trianglelefteq P$ be a normal open subgroup of P. Since P is of finite obliquity P is just infinite. Therefore because P is insoluble there is an $\alpha \in \mathbb{N}$ such that the exponent of $Q/\gamma_2(Q)$ is p^α. Let t be minimal with $A^t = 0$, where A is the augmentation ideal of the group ring of P/Q over $\mathbb{Z}/p^\alpha\mathbb{Z}$. As in the proof of (II.1) one sees that the order of any central section of Q is bounded by p^{kt}. Finally, assume that Q is any open subgroup of P. The core Q_1 of Q is also open in P. So one may assume that any central section of Q_1 has order bounded by p^u for some $u \in \mathbb{N}$. Then any central section of Q is bounded by $|Q : Q_1|p^u$. q.e.d.

(II.5) Proposition. *Let P be a finitely generated infinite pro-p-group. Then P maps onto a just infinite pro-p-group .*

Proof. If P is soluble then P maps onto $\mathbb{Z}_p$ which is just infinite. Hence assume P is insoluble. Let $(N_\alpha)_{\alpha\in\mathbb{N}}$ be an ascending chain of closed normal subgroups of P of infinite index. The lower p-series of some group G is given by $\lambda_{j+1}(G) = [\lambda_j(G), G]G^p$ with $\lambda_1(G) = G$. Define $P_\alpha = P/N_\alpha$. Since P is finitely generated the sequence $(a_j^i)_{j\in\mathbb{N}} := |\lambda_i(P_j) : \lambda_{i+1}(P_j)|$ consists of finite numbers and is constant for $j \geq j(i)$ with certain $j(1) \leq \cdots j(i-1) \leq j(i) \leq \cdots$ for $i \in \mathbb{N}$. Therefore $P_{j(k)}/\lambda_{k+1}(P_{j(k)}) \cong P_{j(k)+n}/\lambda_{k+1}(P_{j(k)+n})$ for all $n \geq 0$ and $k \in \mathbb{N}$. Since one has continuous epimorphisms $P_{j(i)}/\lambda_{i+1}(P_{j(i)}) \to P_{j(i-1)}/\lambda_i(P_{j(i-1)})$ one can define $H = \varprojlim P_{j(i)}/\lambda_{i+1}(P_{j(i)})$ which is an infinite pro-p group. Since the epimorphisms $\varphi_i : P \to P_{j(i)}/\lambda(P_{j(i)})$ are continuous for every $i \in \mathbb{N}$ the epimorphism $\varphi : P \to H$ is continuous. It follows that the closure N of the union of the N_α is contained in Kerφ because Kerφ is closed in P. Since Kerφ is of infinite index in P also N is of infinite index in P. Hence by Zorn's Lemma there exists a maximal normal subgroup M of infinite index in P. Clearly P/M is just infinite. q.e.d.

We finish the chapter with some remarks on wreath products of pro-p-groups with cyclic groups of order p.

(II.6) Lemma. *If G is a just infinite non-abelian pro-p-group then $G \wr C_p$ is also just infinite.*

Proof. The wreath product $G \wr C_p$ equals $(G \times \cdots \times G) \rtimes C_p$, where there are p direct factors, and the action of C_p is by cyclic permutation. Let N be a non-trivial closed normal subgroup of $G \wr C_p$. Let $N_1 = N \cap (G \times \cdots \times G)$. This is a non-trivial subgroup because C_p is not normal in $G \wr C_p$. The map $\pi_i : N_1 \to G$ denotes the projection to the ith factor of $G \times \cdots \times G$. The subgroup $H_i = N_1\pi_i$ is normal and closed in G and therefore of finite index in G for some i. Since G is just infinite and non-abelian the centre $Z(G)$ of G is either trivial or of finite index in G. The latter case is ruled out since by an argument of Schur (cf. [Hup 67] Kapitel V Hilfsatz 23.3 b)) $Z(G) \cap [G,G]$ is a finite normal subgroup of G and hence is trivial. The commutator subgroup $[G,G]$ is of finite index in G and therefore one has a contradiction to G being an infinite group. So $[G,H_i]$ is open in G. Since N_1 is normal in $G \times \cdots \times G$, it follows that the subgroup of $G \times \cdots \times G$ consisting of elements whose i^{th} entry lies in $[G,H_i]$, and whose other entries are the identity, lies in N. Since N is also normalised by a wreathing element, it follows that N contains the direct product of p copies of $[G,H_i]$, and hence is open in $G \wr C_p$. q.e.d.

(II.7) Lemma. *Let G be a pro-p-group with the property that $G/\gamma_2(G)$ is of exponent p. Let $i = ps + t$ where $1 \leq t \leq p$. Let $W = G \wr C_p$. Then $\gamma_i(W)/\gamma_{i+1}(W) \cong \gamma_{s+1}(G)/\gamma_{s+2}(G)$ if $i > 1$, and $W/\gamma_2(W) \cong (G/\gamma_2(G)) \times C_p$.*

Proof. Let B be the base of the wreath product, so that B is the direct product of p copies of G. For $j \geq 0$, $\gamma_j(B)$ is the direct product of p copies of $\gamma_j(G)$, and $\gamma_j(B)/\gamma_{j+1}(B)$ is naturally isomorphic to $\gamma_j(G)/\gamma_{j+1}(G) \otimes \mathbb{F}_p C_p$. Let B_i be the inverse image in B of $\gamma_{s+1}(G)/\gamma_{s+2}(G) \otimes I^{t-1}$, where I is the augmentation ideal of $\mathbb{F}_p C_p$. Note that I^k/I^{k+1} is of order p for $0 \leq k \leq p-1$, and that $I^p = 0$. It is easy to see that, for $i \geq 2$, $B_i = \gamma_i(W)$. For if $i = 2$ then $s = 0$ and $t = 2$, and the result holds in this case, and the general case follows by an easy induction on i. The lemma follows. q.e.d.

If $G/\gamma_2(G)$ is of exponent greater than p the situation is somewhat more complicated. When we meet such cases, we work out the lower central series of the wreath product on an *ad hoc* basis. Note that the Lemma (II.7) implies that the various widths of W are essentially the same as for G.

III p-adically simple groups ($\tilde{p}$-groups)

The fundamental tool for studying these groups is their interconnection with Lie algebras over local fields of characteristic 0.

a) The Baker-Campbell-Hausdorff formula

We thank Bernd Souvignier, cf. [Sou 96], for having worked out the details of this section. To construct groups from Lie algebras, the Baker-Campbell-Hausdorff formula will be used. For some estimations on the p-adic value of its coefficients the following lemma is needed.

(III.1) Lemma. *For $m, n \in \mathbb{N}$, $n = \sum_{i=0}^{s} a_i p^i$ with $0 \le a_i < p$ and $a = \sum_{i=0}^{s} a_i$ the following hold*

(i) $\nu_p(m! \cdot n!) \le \nu_p((m+n)!)$,

(ii) $\nu_p(n!) = \frac{n-a}{p-1}$ *(in particular* $\nu_p(n!) \le \frac{n-1}{p-1}$*).*

Proof. (i) Since $\frac{(m+n)!}{m! \cdot n!} = \binom{m+n}{n} \in \mathbb{N}$, one has $\nu_p((m+n)!) - \nu_p(m! \cdot n!) \ge 0$.
(ii) The number of $a \in \{1, \ldots, n\}$ satisfying $\nu_p(a) \ge i$ is $\lfloor np^{-i} \rfloor$. Therefore the number of a satisfying $\nu_p(a) = i$ is $\lfloor np^{-i} \rfloor - \lfloor np^{-i-1} \rfloor$. Furthermore $\lfloor np^{-i} \rfloor = \sum_{j=0}^{s-i} a_{i+j} p^j$ and it follows that $\nu_p(n!) = \sum_{i=1}^{s} i(\lfloor np^{-i} \rfloor - \lfloor np^{-i-1} \rfloor) = \sum_{i=1}^{s} \lfloor np^{-i} \rfloor = \sum_{i=1}^{s} \sum_{j=0}^{s-i} a_{i+j} p^j = \sum_{k=1}^{s} a_k \sum_{l=0}^{k-1} p^l = \sum_{k=1}^{s} a_k (p^k - 1)/(p-1) = \sum_{k=0}^{s} a_k (p^k - 1)/(p-1) = (n-a)/(p-1)$.
q.e.d.

Let $\mathcal{L}$ be a Lie algebra of finite dimension over a local field of characteristic 0. For $x, y \in \mathcal{L}$ the Baker-Campbell-Hausdorff formula is formally defined as

$$\Phi(x, y) = \log(\exp(x) \cdot \exp(y)).$$

This is a formal definition in the following sense. The definitions of log and exp are taken to be the usual power series definitions. For this to make any sense at all, one needs to work in an associative algebra; of course any Lie algebra can be embedded in an associative Lie algebra, where the Lie product is defined by $[x, y] = xy - yx$. There is also the question of convergence to be considered. By expressing exp and log as power series one has (cf. [Jac 62])

$$\Phi(x, y) = \sum_{m \ge 1} \sum_{\substack{p_i, q_i \\ p_i + q_i > 0}} \frac{(-1)^{m-1}}{m(\sum(p_i + q_i))} \frac{[x_{p_1}, y_{q_1}, \ldots, x_{p_m}, y_{q_m}]}{p_1! q_1! \ldots p_m! q_m!}$$

where $[x_{p_1}, y_{q_1}, \ldots, x_{p_m}, y_{q_m}]$ means $[\underbrace{x, \ldots, x}_{p_1}, \underbrace{y, \ldots, y}_{q_1}, \ldots, \underbrace{x, \ldots, x}_{p_m}, \underbrace{y, \ldots, y}_{q_m}]$. In this expression one has eliminated the associative multiplication, but the problem of convergence remains. Writing $\Phi(x, y) = \sum_{n \ge 1} u_n(x, y)$, where $u_n(x, y)$ is the sum of homogeneous terms of degree n in x, y, yields

$$u_n(x,y) = \sum_{m=1}^{n} \sum_{\substack{p_i, q_i \\ p_i+q_i>0 \\ \Sigma(p_i+q_i)=n}} \frac{(-1)^{m-1}}{mn \cdot p_1!q_1! \ldots p_m!q_m!}[x_{p_1}, y_{q_1}, \ldots, x_{p_m}, y_{q_m}].$$

In particular, $u_1(x,y) = x+y$, $u_2(x,y) = (1/2)[x,y]$, $u_3(x,y) = -(1/12)[x,y,x] + (1/12)[x,y,y]$, $u_4(x,y) = -(1/48)[x,y,x,y] - (1/48)[x,y,y,x]$.

The following table gives, for small n, the least common multiple of the denominators appearing in $u_n(x,y)$; this will be needed for the proof of the next lemma.

n	1	2	3	4	5	6
lcm	1	2	$2^2 \cdot 3$	$2^4 \cdot 3$	$2^4 \cdot 3^2 \cdot 5$	$2^5 \cdot 3^3 \cdot 5$
n	7	8	9	10	11	12
lcm	$2^5 \cdot 3^3 \cdot 5 \cdot 7$	$2^9 \cdot 3^3 \cdot 5 \cdot 7$	$2^8 \cdot 3^5 \cdot 5^2 \cdot 7$	$2^9 \cdot 3^4 \cdot 5^3 \cdot 7$	$2^9 \cdot 3^5 \cdot 5^2 \cdot 7 \cdot 11$	$2^{12} \cdot 3^6 \cdot 5^2 \cdot 7 \cdot 11$

(III.2) Lemma. *Let L be a Lie lattice satisfying $[L,L] \subseteq pL$ if $p \neq 2$, and $[L,L] \subseteq 4L$ if $p = 2$. Then $u_n(x,y) \in L$ for all $x, y \in L$ and all $n \geq 1$. The series $(u_n(x,y))$ converges to $0 \in L$ as $n \to \infty$.*

Proof. From $[L,L] \subseteq pL$ respectively $[L,L] \subseteq 4L$ it follows for the commutator of length s that $[L,\ldots,L] \subseteq p^{s-1}L$ resp. $\subseteq 2^{2s-2}L$. So it is enough to prove for the coefficients $c_n = (-1)^{m-1}(mn \cdot p_1!q_1! \ldots p_m!q_m!)^{-1}$ in $u_n(x,y)$ that $\nu_p(c_n^{-1}) \leq n-1$ respectively $\nu_2(c_n^{-1}) \leq 2(n-1)$.
Clearly by Lemma (III.1), $\nu_p(c_n^{-1}) \leq \nu_p(n!) + \nu_p(n) + \nu_p(m) \leq (n-1)(p-1)^{-1} + 2\log_p(n)$.
(i) $p \geq 5$: Since $\nu_p(m) \leq \nu_p(n!)$ for $m \leq n$, one has $\nu_p(c_n^{-1}) \leq 3\nu_p(n!) \leq 3(p-1)^{-1}(n-1) \leq (p-2)(p-1)^{-1}(n-1)$. Therefore $u_n(x,y) \in p^{(n-1)/(p-1)}L$.
(ii) $p = 3$: For $n \geq 13$ it follows that $n^5 \leq 3^{n-1}$, and therefore $\log_3(n) \leq 5^{-1}(n-1)$. One has $\nu_3(c_n^{-1}) \leq \nu_3(n!) + 2\nu_3(n) \leq (9/10)(n-1)$, hence $u_n(x,y)$ lies in $3^{(n-1)/10}L$. For $n \leq 12$ one checks the claim by using the above table.
(iii) $p = 2$: For $n \geq 13$ one has $n^3 \leq 2^{n-1}$. Then $\log_2(n) \leq 3^{-1}(n-1)$ and $\nu_2(c_n^{-1}) \leq \log_2(n!) + 2\log_2(n) \leq (5/3)(n-1)$. Hence one has $u_n(x,y) \in 2^{(n-1)/3}L$. For $n \leq 12$ one checks the claim by using the above table. q.e.d.

This lemma has the following consequence.

(III.3) Proposition. *For a Lie lattice L satisfying the above conditions $\Phi(x,y)$ is well defined for all $x, y \in L$ and defining $xy = \Phi(x,y)$ makes L into a group.*

Proof. The group axioms follow at once from the formal properties of log and exp. q.e.d.

Note that the zero element of L becomes the identity element in this group, and that the inverse of an element x is $-x$.

To express group commutators in terms of commutators in the Lie lattice one uses the commutator Baker-Campbell-Hausdorff formula defined formally by

$$\Psi(x, y) = \log(\exp(-x)\exp(-y)\exp(x)\exp(y)).$$

By the method used in [Jac 62] one obtains a formula for $\Psi(x, y)$, which is very similar to that of $\Phi(x, y)$. Comparing the p-adic valuations of the coefficients in the homogeneous terms of the same degree in $\Phi(x, y)$ and $\Psi(x, y)$ it is easy to see that the valuations in $\Psi(x, y)$ are greater or equal to those in $\Phi(x, y)$.
Hence one has the following Lemma.

(III.4) Lemma. *Let L be a Lie lattice and X a lattice contained in L such that $[L, X] \subseteq pX$ for p odd, and $[L, X] \subseteq 4X$ if $p = 2$. Then $\Psi(x, y) \in X$ for all $x \in L$ and $y \in X$. Moreover if $X = L$, then $\Psi(x, y) = [x, y]_G$, where $[x, y]_G$ is the commutator of x and y working in the group defined on L with Φ as group multiplication.*

Proof. Again this follows from the formal properties of log and exp. q.e.d.

b) $\tilde{p}$-groups and their Lie algebras

A pro-p-group H is said to be uniformly powerful (or uniform) if it is powerful and satisfies the condition $|\lambda_i(H)/\lambda_{i+1}(H)| = |H/\lambda_2(H)| < \infty$, where $\lambda_i(H)$ denotes the i-th term of the lower p-series of P. Equivalently, H is uniform if H is finitely generated, powerful and torsion free (cf. [DdMS 91] Theorem 4.8.). Using this criterion it is easy to see that the Lie group constructed from a suitable Lie lattice, as in the previous section, is uniform. It is well known that any p-adic analytic group P has an open uniform pro-p-subgroup H, cf. [DdMS 91]. H can be assigned a Lie algebra $L(H)$ over $\mathbb{Z}_p$ in such a way that the assignment P to $\mathcal{L}(P) := \mathbb{Q}_p \otimes L(H)$ defines a functor setting up an equivalence between the category of analytic groups with germs of analytic homomorphisms at 1 as morphisms and the category of $\mathbb{Q}_p$-Lie algebras. The $\mathbb{Z}_p$-Lie algebra $L(H)$, henceforth referred to as the Lie lattice of H, has H itself as the underlying set and carries the following operations expressed in terms of the group operation of H (cf. [Laz 65], or [DdMS 91] part 1). Denote the commutator in the group by [,]. Define an addition on H by

$$g +_L h = \lim_{n\to\infty} (g^{p^n} h^{p^n})^{p^{-n}}$$

and a Lie bracket by

$$[g, h]_L = \lim_{n\to\infty} [g^{p^n}, h^{p^n}]^{p^{-2n}}$$

for all $g, h \in H$. The action of $\mathbb{Z}_p$ is obtained in an obvious way by the powering operation of $\mathbb{Z}$. Note that the conjugation action of H on $L(H)$ is an action by Lie algebra automorphisms.

If H is a uniform pro-p-group, we have defined a $\mathbb{Z}_p$-Lie algebra $L(H)$ with H as its underlying set, and it is easy to see that $[L(H), L(H)] \subseteq pL(H)$ for p odd, and $[L(H), L(H)] \subseteq 4L(H)$ for $p = 2$. It follows that we can then define a group operation on $L(H)$ using the Baker-Campbell-Hausdorff formula. One can check that this gives the original group H, cf. [DdMS 91] Corollary 8.16. Similarly, one can start with a

Lie lattice L, satisfying $[L, L] \subseteq pL$ for p odd, or $[L, L] \subseteq 4L$ for $p = 2$, form the corresponding group using the Baker-Campbell-Hausdorff formula, and then construct the corresponding Lie algebra. This gives back the original Lie algebra.

Clearly the assignment $H \to L(H)$ defines an equivalence between the category of uniform pro-p-groups, and the category of finite dimensional Lie lattices over $\mathbb{Z}_p$. This in turn defines a functor $\mathcal{L}$ from the category of finitely generated p-adic analytic groups to the category of finite dimensional Lie algebras over $\mathbb{Q}_p$ taking P to $\mathcal{L}(P) = L(H) \otimes \mathbb{Q}_p$ where H is an open uniform subgroup of P. It is easy to see that this is a well defined functor, and that the image under the functor of a homomorphism between two finitely generated pro-p-groups depends only on its restriction to an open uniform subgroup. We now investigate the elementary properties of this functor.

Clearly, if H_1 is a uniform subgroup of the uniform group H then $L(H_1)$ is a subalgebra of $L(H)$, and if $H_1 \lhd H$ then $L(H_1) \lhd L(H)$. In the reverse direction, the result is slightly weaker.

(III.5) Lemma. *Let H be a uniform p-adic analytic group and X a Lie sublattice of $L(H)$ satisfying $[X, X]_L \subseteq pX$ if p is odd, and $[X, X]_L \subseteq 4X$ if $p = 2$. Then X, regarded as a subset of H, is a subgroup of H. If, in addition, $[L(H), X]_L \subseteq pX$ if p is odd, and $[L(H), X]_L \subseteq 4X$ if $p = 2$, then X, regarded as a subset of H, is a normal subgroup of H.*

Proof. This follows from the Baker-Campbell-Hausdorff formula, and the commutator Baker-Campbell-Hausdorff formula. q.e.d.

(III.6) Proposition. *Let P be an insoluble $\tilde{p}$-group. Then*

(i) $\mathcal{L}(P)$ is semisimple with isomorphic components, the number of which is a power of p and

(ii) P acts faithfully on $\mathcal{L}(P)$.

Proof. (i) Let I be a characteristic ideal of $\mathcal{L}(P)$ and let H be an open uniform subgroup of P. Then $I \cap L(H)$ is an ideal, J say, in $L(H)$, and if $X = p^2 J$, then X is a normal subgroup of H by Lemma (III.5). Since X is P-invariant, it is a normal subgroup of P. But X is not open in P, since I is a proper ideal of $\mathcal{L}(P)$, so $X = \langle 1 \rangle$ and $I = 0$. P permutes the simple components of $\mathcal{L}(P)$ transitively, since otherwise an orbit of the components under P gives rise to a normal closed subgroup of infinite index. As there is only a finite number of components it has to be a power of p.
(ii) Assume the action of P has a non-trivial kernel. Since P is just infinite it follows that only a finite factor group of P acts non-trivially. Let U be a uniform normal subgroup contained in the kernel of this action. The group P acts trivially by conjugation on the Lie lattice $L(U)$ and therefore it follows that any $g \in P$ and $u \in U$ commute. It follows that U is an abelian open subgroup of P, this contradicts P being insoluble. q.e.d.

Therefore an insoluble $\tilde{p}$-group P will be viewed as a subgroup of the automorphism group $\mathrm{Aut}(\mathcal{L}(P))$ of its Lie algebra $\mathcal{L}(P)$.

c) $\tilde{p}$-groups as open subgroups of groups of automorphisms of Lie algebras

Given a semisimple p-adic Lie algebra $\mathcal{L}$ all derivations are inner, i.e. of the form $\mathrm{ad}l : \mathcal{L} \to \mathcal{L} : \lambda \mapsto [l, \lambda]$ for $l \in \mathcal{L}$, cf. [Zas 39] Satz 15 or 16, or [Jac 62] p. 73. Let $L \subset \mathcal{L}$ be a Lie lattice satisfying $[L, L] \subset p^2 L$. Define $e(l) = \exp(\mathrm{ad}l)$ for any ad-pronilpotent element $l \in L$. Since $(\mathrm{ad}l)^n L \subset p^{2n} L$ and $\nu_p(n!) \leq (n-1)/(p-1)$ it follows that $(1/n!)(\mathrm{ad}l)^n L \subset L$ for all $l \in L$ and $n \in \mathbb{N}$. So, for $l \in L$, $e(l)$ lies in $\mathrm{Aut}(\mathcal{L})$. One uses the Baker-Campbell-Hausdorff formula to see that $e(L)$ is a subgroup, cf. (III.3). Clearly $e(L)$ is a pro-p-subgroup of $\mathrm{Aut}(\mathcal{L})$ ($e(L)$ lies in the centraliser of L/pL) which is closed in $\mathrm{Aut}(\mathcal{L})$ because it is the image of a compactum under a continuous map. The aim is now to construct a $\tilde{p}$-group for a given Lie algebra. This will be achieved in Proposition (III.9).

(III.7) Lemma. *Let $\mathcal{L}$ be a simple finite dimensional $\mathbb{Q}_p$-Lie algebra. Any open pro-p-subgroup P of* $\mathrm{Aut}(\mathcal{L})$ *is a p-adic analytic group.*

Proof. Let n be the dimension of $\mathcal{L}$. The automorphism group $\mathrm{Aut}(\mathcal{L})$ is a closed subgroup of $GL(\mathcal{L}) \cong GL_n(\mathbb{Q}_p)$ which is p-adic analytic. Therefore, by [DdMS 91] Theorem 10.7, $\mathrm{Aut}(\mathcal{L})$ is p-adic analytic itself. Since P is open in $\mathrm{Aut}(\mathcal{L})$ and hence closed the assumption follows by the same argument. q.e.d.

There is another way of constructing a Lie algebra of a p-adic analytic group P. If p is an odd prime, let H be an open uniform subgroup of P. If p is even, let H be the Frattini subgroup of an open uniform subgroup of P. The group algebra $\mathbb{Z}_p H$ can be completed in such a way that a logarithm map log is defined, cf. [DdMS 91] part 2. Set $\Lambda(H) := \log(H)$. Then $\Lambda(H)$ can be given the structure of a $\mathbb{Z}_p$-Lie algebra by defining a Lie bracket employing the product in the group algebra $[x, y] := xy - yx$ and $\Lambda(H)$ is closed under this operation. By using the Baker-Campbell-Hausdorff formula one can check that the map $\log : L(H) \to \Lambda(H) : g \mapsto \log(g)$ is a $\mathbb{Z}_p$-Lie algebra isomorphism.

(III.8) Lemma. *Let $\mathcal{L}$ be a semisimple $\mathbb{Q}_p$-Lie algebra and L a full lattice in $\mathcal{L}$ with $[L, L] \subset p^2 L$. Then $\Lambda(e(L))$ is isomorphic to L via $l \mapsto \log \circ e(l)$.*

Proof. Clearly L is an $e(L)$ - module where the operation is defined by $e(l)(\lambda) = e^{ad(l)}(\lambda) = \sum_{n=0}^{\infty} \frac{1}{n!}(\mathrm{ad}l)^n(\lambda)$ for $l \in L$. The lattice L is also a $\Lambda(e(L))$-module with the operation defined by $\log(e(l))(\lambda) = \sum_{n=1}^{\infty} \frac{(-1)^{n+1}}{n}(1 - e^{\mathrm{ad}l})^n(\lambda) = \mathrm{ad}l(\lambda) = [l, \lambda]$. This is well defined because $(1 - e^{\mathrm{ad}l})^n$ lies in the group algebra $\mathbb{Z}_p[e(L)]$ and is interpreted as acting on the full p-adic lattice L. It follows that the map $\log \circ e$ is an isomorphism of Lie lattices because the action of L defined by this map is faithful. q.e.d.

(III.9) Proposition. *Let $\mathcal{L}$ be a simple finite dimensional $\mathbb{Q}_p$-Lie algebra.*

(i) *Any open pro-p-subgroup P of* $\mathrm{Aut}(\mathcal{L})$ *is a $\tilde{p}$-group and* $\mathcal{L}(P) \cong \mathcal{L}$.

(ii) *Let W be an iterated wreath product of α copies of cyclic groups C_p of order p. An open pro-p-subgroup P of* $\mathrm{Aut}(\mathcal{L}) \wr W$ *is a $\tilde{p}$-group, if and only if it permutes the p^α copies of $\mathcal{L}$ transitively.*

Proof. (i) Applying Lemma (III.7) it remains to prove that there is no closed normal subgroup of infinite index in P.
Choose a Lie lattice $L \subset \mathcal{L}$ and a uniform open subgroup K of P such that $e : L \to K : l \mapsto e^{\mathrm{ad}l}$ is well defined and injective. Furthermore choose L and K in such a way that all open normal subgroups of K are uniform. Then $e(L)$ is normal in K because $ke(l)k^{-1}(\lambda) = k(e^{\mathrm{ad}l}k^{-1}(\lambda)) = e(k(l))(\lambda)$ for all $\lambda, l \in L$, $k \in K$. The Lie lattice $L(e(L))$ is an ideal in $L(K)$ because $[e(l), k] = \lim_{n\to\infty}(e(l)^{p^n}, k^{p^n}) \in e(L)$. Hence the isomorphism of the two constructions of Lie algebras respects subgroups therefore $L \cong \Lambda(e(L)) \triangleleft \Lambda(K)$ and consequently $\mathcal{L} \triangleleft \mathcal{L}(P)$. Since $\mathcal{L}$ is simple there are no outer derivations. Thus $\mathcal{L}(P) \cong \mathcal{L} \oplus \mathcal{X}$ decomposed as Lie algebras. It remains to show that $\mathcal{X} = \{0\}$. Choose K and L such that $\Lambda(K) = L \oplus X$ as Lie lattices. The action of K on L is faithful. The element $e(x) \in K$ for $x \in X$ acts trivially on L. Therefore $X = \{0\}$.
(ii) To prove that P is just infinite under the assumed condition, see the proof of the first claim of Lemma (II.6). Conversely, let $L(U)$ be a Lie lattice associated to P which decomposes into $\oplus_i C_i$. A non-transitive action of P on $L(U)$ yields an orbit of infinite index in $L(U)$ and therefore corresponds to a closed normal subgroup in P of infinite index which yields a contradiction to P being just infinite. q.e.d.

(III.10) Corollary. *An insoluble $\tilde{p}$-group P is an open subgroup of* $\mathrm{Aut}(\mathcal{L}) \wr W$*, where $\mathcal{L}$ is a simple Lie algebra over $\mathbb{Q}_p$ and W is an iterated wreath product of α copies of cyclic groups C_p of order p such that* $\mathcal{L}(P) \cong \oplus^{p^\alpha} \mathcal{L}$.

Proof. It follows from (III.6) that P is a closed subgroup of $\mathrm{Aut}(\mathcal{L}(P))$. Denote the dimension of $\mathcal{L}(P)$ by d. Let G be an open subgroup of $\mathrm{Aut}(\mathcal{L}(P))$ which is a $\tilde{p}$-group. Take P_1 to be an open uniform subgroup of P and G_1 to be an open uniform subgroup of G. Then $L(P_1)$ and $L(G_1)$ are both full Lie lattices in $\mathcal{L}(P)$ of dimension d and hence $L(P_1) \cap L(G_1)$ is a full Lie lattice in $\mathcal{L}(P)$. Therefore $P_1 \cap G_1$ is of finite index in G and P is open in $\mathrm{Aut}(\mathcal{L}(P))$. The rest of the claim follows from Proposition (III.6). q.e.d.

(III.11) Lemma. *An insoluble $\tilde{p}$-group P is not isomorphic to one of its proper subgroups.*

Proof. Assume P_1 to be a subgroup of P isomorphic to P. Let H be a uniform normal subgroup of P and H_1 the subgroup of P_1 corresponding to H under the isomorphism from P to P_1. The Lie lattices $L(H)$ and $L(H_1)$ are isomorphic and $L(H_1)$ is contained in $L(H)$. The Killing form on $L(H)$ and $L(H_1)$ is not degenerate and has the same discriminant on both lattices. Therefore these Lie lattices are equal. It follows that the subgroups H and H_1 are equal because they are equal as sets. The finite factor groups P/H and P_1/H are of the same order, so $P_1 = P$. q.e.d.

This result is in contrast to the situation of linear groups of finite width over $\mathbb{F}_p[[t]]$. For instance the Sylow pro-p-subgroup of $SL_n(\mathbb{F}_p[[t]])$ is isomorphic to the Sylow pro-p-subgroup $SL_n(\mathbb{F}_p[[t^2]])$ which it contains properly as a subgroup of infinite index. The Nottingham group also contains proper subgroups isomorphic to itself. It may be that an insoluble just infinite pro-p-group P contains no proper isomorphic subgroup isomorphic to itself if and only if P is p-adic analytic.

d) Obliquity and lattices of normal subgroups

The following theorem shows that there is a strong link between the uniform normal subgroups of an insoluble $\tilde{p}$-group P and the full Lie lattices in its Lie algebra $\mathcal{L}(P)$. This almost determines the structure of the lattice of normal subgroups of P since there exists an open uniform characteristic normal subgroup T such that all normal subgroups of P which are contained in T are uniform, cf. [DdMS 91] Corollary 4.5.

(III.12) Theorem. *Let P be an insoluble $\tilde{p}$-group with $\mathbb{Q}_p$-Lie algebra $\mathcal{L}(P)$. Then there are finitely many P-invariant Lie lattices $L_1, \ldots, L_k$ in $\mathcal{L}(P)$ such that the following holds.*

(i) Every P-invariant Lie lattice in $\mathcal{L}(P)$ is of the form $p^\alpha L_i$ for some $i, \alpha \in \mathbb{N}$.

(ii) There exist integers $\alpha_1, \ldots, \alpha_k$ such that there is a bijection between the sets $\{N \lhd P | N \text{ uniform in } P\}$ and $\{p^\alpha L_i | \alpha \in \mathbb{Z}_{\geq 0}, \alpha \geq \alpha_i, 1 \leq i \leq k\}$ via $N \mapsto L(N)$.

Proof. (i) P acts irreducibly on $\mathcal{L}(P)$. Therefore, by the Jordan-Zassenhaus-Theorem, one has only finitely many isomorphism classes of P-invariant lattices in $\mathcal{L}(P)$.
(ii) Define the set $\mathcal{N} := \{N \lhd P | N \text{ uniform in } P\}$. There are $N_1, \ldots, N_l \in \mathcal{N}$ which are maximal with respect to the condition that there exists no element $M \in \mathcal{N}$ such that $M^{p^i} = N$ for any $i \in \mathbb{N}$. It follows from (i) that $l = k$ and $L(N_i) = p^{\alpha_i} L_i$ since $L(N_i) \not\subset p^m L(N_j)$ for any $1 \leq i, j \leq k$, $m \in \mathbb{N}$. q.e.d.

(III.13) Corollary. *Let P be an insoluble $\tilde{p}$-group. Then P is ultimately periodic in $|\gamma_i(P) : \gamma_{i+1}(P)|$.*

Moreover one has the following.

(III.14) Corollary. *Let P be an insoluble $\tilde{p}$-group. Then there are numbers $n_0, z, f \in \mathbb{N}$ such that for any $n \geq n_0$ one has $\gamma_{n+z}(P) = \mathcal{F}^f(\gamma_n(P)) = \gamma_n(P)^{p^f}$, where $\mathcal{F}(X) = \mathcal{F}^1(X)$ is the Frattini subgroup of the group X and $\mathcal{F}^{i+1}(X) := \mathcal{F}(\mathcal{F}^i(X))$.*

Call the lexicographic smallest such pair (z, f) the periodicity of P and this f the defect of the periodicity. The existence of maximal $\tilde{p}$-groups will be established in the next section. The maximal $\tilde{p}$-groups which we have seen so far had 1 as their defect of periodicity. It is clear that this defect can become arbitrarily large, if the index of P in its maximal $\tilde{p}$-group gets large. Note, the average width is related to these parameters by

$$w_a(P) = d \cdot \frac{f}{z}$$

where d is the dimension of the Lie algebra $\mathcal{L}(P)$.

Theorem (III.12) indicates that the lattice of normal subgroups of P is essentially determined by a sufficiently big finite factor group. The simplest possibility is that each open normal subgroup of P lies between some $\gamma_i(P)$ and $\gamma_{i+1}(P)$. The deviation from this behaviour is measured by the obliquity. Recall its definition.

(III.15) Definition. *Let P be a pro-p-group of finite width. Define*

$$\mu_i(P) := \Big(\bigcap_{N \not\subset \gamma_{i+1}(P), N \triangleleft P} N \Big) \cap \gamma_{i+1}(P).$$

Then $o(P) := max_{i \in \mathbb{N}} \log_p(|\gamma_{i+1}(P) : \mu_i(P)|)$ *if it exists or otherwise* $o(P) := \infty$ *is called the obliquity of P.*

If P is an insoluble $\tilde{p}$-group then by Theorem (III.12) and since P is just infinite $o(P)$ is well defined and finite. Using the correspondence between the open normal subgroups and the Lie lattices one can determine the ultimate obliquity $o_u(P) := \overline{\lim}_{i \to \infty} \log_p(|\gamma_{i+1}(P) : \mu_i(P)|)$ from the Lie sublattices.

e) Uniqueness and existence of maximal $\tilde{p}$-groups

Call a $\tilde{p}$-group maximal, if it is not properly contained in another $\tilde{p}$-group as an open subgroup. Let $\mathcal{L}$ be a semisimple finite dimensional $\mathbb{Q}_p$-Lie algebra. One can view $\mathrm{Aut}(\mathcal{L}) = A(\mathbb{Q}_p)$ as the group of $\mathbb{Q}_p$-rational points of a semisimple algebraic group A. Let $\mathrm{Aut}(\mathcal{L})^\circ := A^\circ(\mathbb{Q}_p)$ where A° is the connected component of 1 in A. One can apply to $\mathrm{Aut}(\mathcal{L})^\circ$ an analogue of the Sylow theorem by Matsumoto, cf. [Mat 66]. This proves that $\mathrm{Aut}(\mathcal{L})^\circ$ contains an open maximal pro-p-group. Any two open maximal pro-p-groups are conjugate in $\mathrm{Aut}(\mathcal{L})$. Furthermore any pro-$p$-group is contained in an open maximal pro-p-group. It is well known that the index $|A : A^\circ|$ is finite. Furthermore A° is normal in A. Therefore, to get a Sylow theory for $\mathrm{Aut}(\mathcal{L})$, one has to deal with finite extensions of a connected semisimple algebraic group. This is done in the following lemma.

(III.16) Lemma. *Every pro-p-subgroup of* $\mathrm{Aut}(\mathcal{L})$ *is contained in an open maximal pro-p-subgroup. All open maximal pro-p-subgroups are conjugate in* $\mathrm{Aut}(\mathcal{L})$.

Proof. Let $G^\circ = \mathrm{Aut}^\circ(\mathcal{L})$ and $G = \mathrm{Aut}(\mathcal{L})$. Since G° has finite index in G and is normal in G one can deduce a Sylow theorem for G as follows.
For the first step let P be an open pro-p-subgroup in G°. From [Mat 66] Proposition 1, it follows that the normaliser N_{G° of an open pro-p-subgroup P is compact because P itself is compact. Therefore an open (Sylow) pro-p-subgroup P in G° is contained in its normaliser $N_{G^\circ}(P)$ with finite (p-prime) index. For any $x \in G$ there exists

$g \in G^\circ$ such that $P^x = P^g$. In particular xg^{-1} lies in $N_G(P)$. Therefore every element x in G can be written as a product of an element in $N_G(P)$ and an element in G°. It follows that the factor $N_G(P)/N_{G^\circ}(P)$ is isomorphic to the finite group G/G°. Applying Sylow's theorem to the finite quotient $N_G(P)/P$ the pre-image $\tilde{P}$ of a Sylow p-subgroup of $N_G(P)/P$ is a maximal pro-p-subgroup of G. Clearly any two maximal pro-p-subgroups of G obtained in this way are conjugate. Let Q be an open maximal pro-p-subgroup of G. Claim, Q is conjugate to $\tilde{P}$ under G. To prove this let $Y = G^\circ \cap Q$. Then $|N_{G^\circ}(Y) : Y|$ is finite and Q normalises $N_{G^\circ}(Y)$. Hence $N_{G^\circ}(Y)Q$ is an open subgroup of G with Y as an open normal subgroup. Since Q is a maximal pro-p-subgroup in G the prime $p \not| |N_{G^\circ}(Y)Q : Q| = |N_{G^\circ}(Y) : Y|$, hence Y is a maximal pro-p-subgroup of G°, proving the claim.

It remains to deal with the case that the maximal pro-p-subgroup Q is not open in G. As proved in [Mat 66] one has that $Y := Q \cap G^o$ lies in some open pro-p-Sylow subgroup P of G°. The intersection $\bigcap_{g \in Q} P^g$ is of finite index in P and therefore open. Then Q lies in the open pro-p-subgroup $(\bigcap_{g \in Q} P^g)Q$ and by maximality it must be equal to it, contradicting the assumption that Q is not open. q.e.d.

(III.17) Corollary. *An insoluble $\tilde{p}$-group P embeds into a maximal $\tilde{p}$-group P_{max} with the same Lie algebra $\mathcal{L}(P) = \mathcal{L}(P_{max})$. Any two groups P_{max} with this property are isomorphic. The group P_{max} can be obtained by constructing a series of open subnormal subgroups $P_1 = P \lhd P_2 \lhd \cdots \lhd P_{max}$ in the following way: P_i is the Sylow pro-p-subgroup of* $\mathrm{Aut}(P_{i-1})$.

Proof. Clearly P lies in a Sylow pro-p-subgroup of $\mathrm{Aut}(\mathcal{L}(P))$. Note, $\mathrm{Aut}(\mathcal{L}(P))$ acts faithfully on $\mathcal{L}(P)$ and the normaliser of a proper open subgroup X of a pro-p-group contains X properly. q.e.d.

The dimension of an analytic pro-p-group is the dimension of its Lie algebra.

(III.18) Proposition. *There are only finitely many maximal $\tilde{p}$-groups of given dimension.*

Proof. Namely for a given finite extension K of $\mathbb{Q}_p$ there are only finitely many absolutely simple Lie algebras over K of given dimension ([Kne 65], [Sat 71] p. 119). There are only a finite number of extensions $K/\mathbb{Q}_p$ of given degree, cf. [Nar 90] p. 216. q.e.d.

IV Periodicity

As examples of the kind of result we have in mind, consider two of N. Blackburn's theorems. The first states that if P is a p-group such that $P/\gamma_{p+1}(P)$ is of maximal class then so is P (cf. [Bla 58]). The second states that if P is a finite p-group with $p > 3$, $|P/\gamma_3(P)| = p^3$, and $P^p = \gamma_3(P)$, defining $f_i = \log_p|\gamma_i(P)/\gamma_{i+1}(P)|$ then $(f_1, f_2, \ldots) = (2,1,2,1,\ldots,1,2,1,f)$, with $0 \leq f \leq 2$ (cf. [Hup 67] p. 392, [Bla 61]). Also $[\gamma_i(P), \gamma_j(P)] = \gamma_{i+j}(P)$ if i or j is odd. The former result has been generalised in [Lee 94b] and [Sha 94] for sufficiently large p-groups of given coclass. The latter is much easier to generalise. We have not stated our generalisation in the strongest possible form, so it is easy, using the same idea, to obtain a stronger result in various special cases. In fact our 'generalisation' yields a weaker result, as it stands, when restricted to Blackburn's original situation, than that obtained by Blackburn.

(IV.1) Definition. *Let P be a finite p-group of class c, and let $n \geq 1$. Let $N = \gamma_n(P)$ if p is odd, let $N = \gamma_n(P)^2$ if $p = 2$. Denote by $d(N)$ the number of generators of N. Then P is settled with respect to n if*

(i) $d(N) \leq n$, and

(ii) $N^p \geq \gamma_c(P)$.

We should observe that this is not the same concept as that used in [Lee 94b]. However the concepts are similar, in that in both cases, the structure of a pro-p-group is largely determined by its quotient by some term of the lower central series provided that the quotient is settled.

(IV.2) Remark. By Proposition 1.13 and 4.1.13 of [LuM 87a], it follows that N is strongly hereditarily powerful in P. This means that if $M \lhd P$ and $M \leq N$, then M is powerfully embedded in N; that is to say, $[M, N] \leq M^p$ if p is odd, and $[M, N] \leq \mho_2(M)$ if $p = 2$, when $\mho_1(H) := H^p$ and $\mho_{i+1}(H) = \mho_1(\mho_i(H))$ for $i \geq 1$.

(IV.3) Lemma. *Let P be a finite p-group of class c such that $P/\gamma_c(P)$ is settled with respect to n. Then P is settled with respect to n.*

Proof. Assume that p is odd. Then $\gamma_n(P/\gamma_c(P)) = \gamma_n(P)/\gamma_c(P)$. However $(\gamma_n(P/\gamma_c(P)))^p \geq \gamma_{c-1}(P)/\gamma_c(P)$, so $d(\gamma_n(P/\gamma_c(P))) = d(\gamma_n(P))$. The result follows. The proof for $p = 2$ is similar. q.e.d.

(IV.4) Proposition. *Let N be a strongly hereditarily powerful subgroup of the p-group P. Then $\pi : M \mapsto M^p$ defines an isomorphism between the lattice $\mathcal{L}_1$ of normal subgroups K of P such that $\Omega(N) \leq K \leq N$, and the lattice $\mathcal{L}_2$ of normal subgroups L of P satisfying $\langle 1 \rangle \leq L \leq N^p$ with $\Omega(N) = \langle n \in N | n^p = 1 \rangle$.*

Proof. Let p be odd. We first prove that π is onto. Let $L \in \mathcal{L}_2$. By Proposition 1.7 of [LuM 87a] every element of N^p can be written as a pth-power of an element in N since N is powerful. Hence $L = \{g^p : g \in N, g^p \in L\}$. Let $K = \{g \in N : g^p \in L\}$.

Since L is powerfully embedded in N, it follows that K is a subgroup of N, and clearly $K \in \mathcal{L}_1$, and $\pi(K) = L$.
We now prove that π is injective. Let $M \in \mathcal{L}_1$, and $M^p = L \in \mathcal{L}_2$. Let $x^p \in L$. We prove, by induction on $|L|$, that $x \in M$. This is trivial if $L = \langle 1 \rangle$. By Proposition 1.7 of [LuM 87a], $\exists y \in M : y^p = x^p$. Let $x = yz$, so $z \in N$. Then $z^p \in [M,N]^p$ as M is powerfully embedded in N, and $[M,N]^p = [M^p, N]$ by the remark to Proposition 1.6 of [LuM 87a]. So by induction, $z \in [M,N]\Omega(N)$. Hence $z \in M$ and $x \in M$, as required. It follows that $M = \{x \in N : x^p \in L\}$, so M is determined by L.
As π preserves inclusions, it follows that π is a lattice isomorphism.
The proof for $p = 2$ is similar. q.e.d.

(IV.5) Definition. *Let P be an infinite pro-p-group, and let $n \geq 1$. Then P is settled with respect to n if, for some c, we have $P/\gamma_{c+1}(P)$ is settled with respect to n.*

(IV.6) Proposition. *Let P be an infinite pro-p-group that is settled with respect to n. Let $N = \gamma_n(P)$ if p is odd, and let $N = \gamma_n(P)^2$ if $p = 2$. Then N is a strongly hereditarily powerful subgroup of P, and $\pi : M \mapsto M^p$ defines an isomorphism between the lattice $\mathcal{L}_1$ of closed normal subgroups K of P such that $\Omega(N) \leq K \leq N$, and the lattice $\mathcal{L}_2$ of closed normal subgroups L of P satisfying $\langle 1 \rangle \leq L \leq N^p$ with $\Omega(N) = \langle n \in N | n^p = 1 \rangle$.*

Proof. This follows from Lemma (IV.3) and Proposition (IV.4), by considering suitable quotients of P. q.e.d.

(IV.7) Proposition. *If P is an infinite pro-p-group that is settled with respect to n, then P is p-adic analytic. Also, if P is just infinite, then $\gamma_n(P)$ is uniform if p is odd; and $\gamma_n(P)^2$ is uniform if $p = 2$.*

Proof. The first statement holds because P contains a powerful subgroup $N = \gamma_n(P)$ if p is odd or $N = \gamma_n(P)^2$ if p is even. To prove that N is uniform it is sufficient to prove, by Theorem 4.8 [DdMS 91], that N is torsion free. If not then $\Omega(P)$ is a non-trivial closed normal subgroup of P, and hence is open. But this is impossible by Proposition (IV.6). q.e.d.

(IV.8) Lemma. *Let P be a p-adic analytic group, and let N be a uniform subgroup of P. Then $L(M^p) = pL(M)$.*

Proof. Trivial. q.e.d.

(IV.9) Lemma. *Let $M \leq N$ where N is a strongly hereditarily powerful subgroup of the p-adic analytic group P. Then the set of $x, y \in N, x \equiv y \bmod M$ has the same content if the congruence is interpreted group-theoretically or Lie-theoretically.*

Proof. If $xy^{-1} \in M$ then $x - y = \lim(x^{p^i}y^{-p^i})^{p^{-i}} = \lim z_i$, where $\{z_i\}$ is a convergent sequence of elements in M, since $[M,N]$ is powerfully embedded in N. Conversely, if $x - y \in M$ then since $(x^{p^i}y^{-p^i})^{p^{-i}} \equiv xy^{-1} \bmod [M,N]$, it follows that $xy^{-1} \in M$. q.e.d.

Before proving the next result about these subgroups, we need a commutator formula.

(IV.10) Lemma. *Let G be a group, and $x, y \in G$. Let p be a prime, and $i > 0$. Then $y^{p^i}x = xy^{p^i}[y,x]^{p^i}c_0c_1\cdots c_i$, where for all j, the term c_j is a product of p^{i-j}-th powers of basic commutators in x and y involving at least p^j copies of y for $j > 0$, and at least two copies for $j = 0$.*

Proof. Collect the word $y^{p^i}x$ by collection from the left. That is to say, always collect the left-most term that is not in place. To keep track of the collection process, label each of the original occurrences of y as $y_1, y_2, \ldots, y_{p^i}$. That is to say, we proceed as if we had p^i distinct commuting elements. Say that two basic commutators are equivalent if they become identical when suffices are ignored, and all equalities and inequalities between suffices in corresponding places are the same in both words. So, for example $[[y_3, x, y_1, y_2, y_4], [y_2, x]]$ is equivalent to $[[y_4, x, y_2, y_3, y_6], [y_3, x]]$. Now if a basic commutator, with suffices, occurs in the collected word, it occurs only once, and every equivalent word also occurs. The number of commutators in this equivalence class is $\binom{p^i}{k}$, where k is the number of distinct suffices that occur in the original commutator expression. But $\binom{p^i}{k}$ is a multiple of p^j where p divides k exactly $i-j$ times, for $k \leq p^i$. This gives the required formula, once one has observed that $[y,x]$ occurs exactly p^i times in the collection. q.e.d.

(IV.11) Lemma. *Let P be a p-adic-analytic group, and N be a strongly hereditarily powerful uniform normal subgroup of P. Then $n - n^g \equiv [n,g] \bmod [N,P]^p$. By Lemma (IV.9), this is unambiguous.*

Proof. By Lemma (IV.10), $n - n^g$ is the limit of the convergent sequence $\{y_i\}$ where $y_i = ((n^{p^i}(n^g)^{-p^i})^{p^{-i}} = (n^{p^i}gn^{-p^i}g^{-1})^{p^{-i}} = ([n,g]^{p^i}x_i^{p^i})^{p^{-i}}$ where $x_i = c_{0i}c_{1i}\cdots c_{ii}$. Here $c_{ji} \in [N,P]^p$ for all j, since N is a strongly hereditarily powerful subgroup of N. Since N is powerful, $[n,g]^{p^i}x_i^{p^i} = ([n,g]m_i)^{p^i}$ for some m_i in N. Since $m_i^{p^i} \in [N,P]^{p^{i+1}}$, it follows that $m_i \in [N,P]^p$ since N is uniform. So $n - n^g$ is the limit of the convergent sequence $[n,g]m_i$, where $m_i \in [N,P]^p$, and this completes the proof. q.e.d.

(IV.12) Proposition. *Let P be a p-adic analytic group, and let N be a strongly hereditarily powerful uniform normal subgroup of P. Then the Lie ideal $[L(N), P]$ and the normal subgroup $[N,P]$ are equal.*

Proof. This follows from the previous result. q.e.d.

(IV.13) Proposition. *Let M and N be strongly hereditarily powerful subgroups of the p-adic analytic group P. If $L(M)$ and $L(N)$ are isomorphic as P-modules, then $M/[M,P]$ and $N/[N,P]$ are isomorphic as groups.*

Proof. By the previous result, $[M,P]$ and $[N,P]$ may be interpreted in the Lie sense, so $M/[M,P]$ and $N/[N,P]$ are isomorphic, when both sides are interpreted as abelian groups with addition induced by the Lie addition. But as these quotients are abelian, this is simply the addition induced by the group multiplication. q.e.d.

The three theorems with which we end this section give our 'generalisation' of the theorem of Blackburn with which the section began.

(IV.14) Theorem. *Let $\gamma_i(P)$ be a strongly hereditarily powerful subgroup of the p-adic analytic group P. Then there is a $t > 0$ such that*

$$\gamma_j(P)/\gamma_{j+1}(P) \cong \gamma_{j+t}(P)/\gamma_{j+t+1}(P)$$

for all $j \geq i$.

Proof. The sequence $\{L(\gamma_j(P)) : j \geq i\}$ contains only finitely many equivalence classes, where M and N are defined to be equivalent if $M = p^r N$ for some (not necessarily positive) integer r. So for some j and $t > 0$ we have $\gamma_{j+t}(P) = \gamma_j(P)^{p^r}$. Then by Lemma (IV.8) and Proposition (IV.11), $\gamma_{j+t+1}(P) = \gamma_{j+1}(P)^{p^r}$. It follows that $\gamma_{k+t}(P) = \gamma_k(P)^{p^r}$ for all $k \geq j$. As this can be interpreted as a statement about the Lie lattice, we can deduce that $\gamma_{j+t-1}(P) = \gamma_{j-1}(P)^{p^r}$ provided that $j > i$. Iterating gives $\gamma_{i+t}(P) = \gamma_i(P)^{p^r}$. It follows that $L(\gamma_i(P))$ and $L(\gamma_{i+t}(P))$ are isomorphic as P-modules. The result now follows from Proposition (IV.13). q.e.d.

(IV.15) Theorem. *Let P be a finite or infinite pro-p-group such that $P/\gamma_{c+1}(P)$ is settled with respect to n for some $n \geq 2$. If $P/\gamma_{c+1}(P) \cong Q/\gamma_{c+1}(Q)$ for some just infinite pro-p-group Q, and $\gamma_n(Q)^p = \gamma_{n+k}(Q)$ for some $k > 0$, then $\gamma_i(P)/\gamma_{i+1}(P)$ is a homomorphic image of $\gamma_i(Q)/\gamma_{i+1}(Q)$ for all $i \geq 1$.*

Proof. Define a set-theoretic map $\theta : Q \to P$ as follows. Let $\alpha : Q/\gamma_{c+1}(Q) \to P/\gamma_{c+1}(P)$ be an isomorphism. Fix a transversal $\{g_1, \ldots, g_r\}$ of $Q/\gamma_{c+1}(Q)$ and define $g_i\theta$ to be a pre-image in P of $\alpha(g_i\gamma_{c+1}(Q))$. For $h \in Q - \gamma_n(Q)^p$ with $h\gamma_{c+1}(Q) = g_i\gamma_{c+1}(Q)$ define $h\theta := g_i\theta$. Now define inductively $\theta : \gamma_n(Q)^{p^a} - \gamma_n(Q)^{p^{a+1}} \to P$ for $a = 1, 2, \ldots$ by $g\theta = (g^{1/p}\theta)^p$. Clearly θ maps $\gamma_i(Q)$ onto $\gamma_i(P)$ for all i, and induces a surjection of $\gamma_i(Q)/\gamma_{i+1}(Q)$ onto $\gamma_i(P)/\gamma_{i+1}(P)$ for all i. q.e.d.

(IV.16) Remark. Although the condition that $\gamma_n(Q)^p = \gamma_{n+k}(Q)$ for some $k > 0$ is satisfied in most interesting cases, if it is not, we can replace it by an alternative condition that is always satisfied. We can simply assume that if N is any normal subgroup of Q such that $N \leq \gamma_n(Q)$ but $N \not\leq \gamma_n(Q)^p$ then $N \geq \gamma_c(Q)$. Note that this condition is satisfied for c big enough. In particular c depends on the obliquity of Q.

(IV.17) Theorem. *Let P be a pro-p-group. Let $P/\gamma_{c+1}(P) \cong Q/\gamma_{c+1}(Q)$ for some just infinite pro-p-group Q, where the length of the period of the sequence $g_i := \log_p(|\gamma_i(Q) : \gamma_{i+1}(Q)|)$ is k. Let $P/\gamma_{c+1}(P)$ be settled with respect to n for some $n \geq 2$, and c sufficiently big. Define $f_i := \log_p(|\gamma_i(P) : \gamma_{i+1}(P)|)$. If $f_i < g_i$ for some $i \geq c+1$ then P is finite. The size of c depends on the obliquity. Define $m_i := \min\{j \in \mathbb{N} \mid \gamma_{m_j} \leq \mu_i(P)\} - (i+1)$ and $m := \max\{m_i \mid i \in \mathbb{N}\}$, for the definition of $\mu_i(P)$ see (III.15). If $\gamma_n(Q)^p = \gamma_{n+k}(Q)$ assume $c+1 \geq n+m+k$ otherwise use an alternative condition as in Remark (IV.16).*

Proof. Let N be the set of elements of Q mapped to the identity by θ, where θ is as in the proof of the last result. It is easy to see that N is a closed normal subgroup of Q, and that the image of $\gamma_i(Q)/\gamma_{i+1}(Q)$ under θ is isomorphic to $\gamma_i(Q)N/\gamma_{i+1}(Q)N$. The result follows. q.e.d.

V Chevalley groups

To get a rough idea how the $\tilde{p}$-groups and their lower central series look like, we investigate the Sylow pro-p-subgroups of Chevalley groups $G = G(\Phi, K)$ of adjoint type, where Φ is a simple root system and K is a local field with finite residue class field F of characteristic p. These groups are always just infinite of finite width. In the case that the characteristic of K is zero, they are $\tilde{p}$-groups and often they are even maximal $\tilde{p}$-groups.

Notation

Φ a simple root system with base Δ,
$\Phi^i := \{\beta = \sum_{\alpha\in\Delta} a_\alpha \alpha | \beta \in \Phi, a_\alpha \in \mathbb{Z}, \mathrm{ht}(\beta) := \sum_{\alpha\in\Delta} a_\alpha = i\}$ the set of roots of height i. So $\Phi^1 = \Delta$ and $|\Phi^k| = 1$ for the smallest $k \in \mathbb{Z}$ with $\Phi^k \neq \emptyset$, since there is a unique highest root.
Φ^+ the set of all positive roots and Φ^- the set of all negative roots with respect to the base Δ,
Q the weight lattice of Φ (note, $|Q : \mathbb{Z}\Phi|$ is equal to the determinant of the Cartan matrix),
$\mathcal{L}$ a complex Lie algebra with Chevalley basis $(H_\alpha, X_\beta | \ \alpha \in \Delta,\ \beta \in \Phi)$ cf. [Ste 68],
$\langle \alpha, \beta \rangle (= 2(\alpha,\beta)/(\beta,\beta))$ the Cartan numbers for roots $\alpha, \beta \in \Phi$,
$\mathcal{L}(\mathbb{Z})$ the Chevalley lattice, i.e. the $\mathbb{Z}$-sublattice of $\mathcal{L}$ spanned by the Chevalley basis,
$\mathcal{L}(R) := R \otimes \mathcal{L}(\mathbb{Z})$ for any commutative ring R,
K a local field (i.e. a field which is complete with respect to a discrete valuation $\nu : K \to \mathbb{Z} \cup \{\infty\}$) whose valuation ring $\mathcal{O}$ has finite residue class field $F = \mathcal{O}/\pi\mathcal{O}$ with π a uniformising element of $\mathcal{O}$,
$\bar{X}_\alpha$ and $\bar{H}_\alpha$ the images of X_α and H_α in $\mathcal{L}(K)$ respectively $\mathcal{L}(\mathcal{O})$.

As shown by Chevalley, cf. [Che 55] (see also [Car 72], [Ste 68]), there is a homomorphism $\varphi_\alpha : SL_2(K) \to \mathrm{Aut}(\mathcal{L}(K))$ for each root $\alpha \in \Phi$. Write

$$x_\alpha(t) := \varphi_\alpha(\begin{pmatrix} 1 & t \\ 0 & 1 \end{pmatrix})(= \exp(t\bar{X}_\alpha)).$$

Then

$$x_{-\alpha}(t) = \varphi_\alpha(\begin{pmatrix} 1 & 0 \\ t & 1 \end{pmatrix}).$$

Define $G(\Phi, K) := \langle x_\alpha(t) | \alpha \in \Phi, t \in K \rangle$ to be the Chevalley group of type $\mathcal{L}$ over K. Set

$$\mathcal{X}_{\alpha,r} = \{x_\alpha(t) | \ t \in \mathcal{O}, \nu(t) \geq r\}$$

for $\alpha \in \Phi$ and $r \in \mathbb{N}$ and

$$\mathcal{H}_r = \{h(\chi) | \ \chi \in \mathrm{Hom}(\mathbb{Z}\Phi, \mathcal{O}_r^*)\}$$

with $\mathcal{O}_r^* = \{x \in \mathcal{O} | \ \nu(1-x) \geq r\}$ where $h(\chi)\bar{H}_\alpha = \bar{H}_\alpha$ and $h(\chi)\bar{X}_\alpha = \chi(\alpha)\bar{X}_\alpha$ for all $\alpha \in \Phi$. Then $\mathcal{X}_{\alpha,r}$ and $\mathcal{H}_r$ are subgroups of $\mathrm{Aut}(\mathcal{L}(K))$, even of $\mathrm{Aut}(\mathcal{L}(\mathcal{O}))$. We want to study the lower central series of

$$P = P(\Phi, \mathcal{O}) := \langle \mathcal{X}_{\alpha,0},\ \mathcal{X}_{\beta,1},\ \mathcal{H}_1 \mid \alpha \in \Phi^+, \beta \in \Phi^- \rangle.$$

(V.1) Remark. *The group $P(\Phi, \mathcal{O})$ is a Sylow pro-p-subgroup of* $\mathrm{Aut}(\mathcal{L}(K))$ *for* $char(K) = 0$ *unless* $p \mid |Q : \mathbb{Z}\Phi|$ *or* $p \mid |\mathrm{Aut}(K/\mathbb{Q}_p)|$ *or there is a diagram automorphism of the Dynkin diagram of* Φ *of order* p.

For a proof see [Don 87] p. 327-330 and [Iwa 66].

To see the lower central series of P it is easier to deal with its Lie algebra analogue first. Let

$$\begin{aligned} \mathcal{L}_j = \mathcal{L}_j(\mathcal{O}) \quad = \quad & \bigoplus_{i=1}^{j-1} \Big(\bigoplus_{\alpha \in \Phi^i} \pi \mathcal{O} \bar{X}_\alpha \oplus \bigoplus_{\beta \in \Phi^{-k-1+i}} \pi^2 \mathcal{O} \bar{X}_\beta \Big) \oplus \\ & \bigoplus_{i=j}^{k} \Big(\bigoplus_{\gamma \in \Phi^i} \mathcal{O} \bar{X}_\gamma \oplus \bigoplus_{\delta \in \Phi^{-k-1+i}} \pi \mathcal{O} \bar{X}_\delta \Big) \oplus \\ & \bigoplus_{\alpha \in \Phi^1} \pi \mathcal{O} \bar{H}_\alpha \end{aligned}$$

for $j = 1, \ldots, k+1$, where k is the maximal height of the roots in Φ. For $j = (k+1)q + r$ with $q \geq 0$ and $1 \leq r \leq k+1$ let $\mathcal{L}_j = \pi^q \mathcal{O} \mathcal{L}_r$. Then the $\mathcal{L}_j(\mathcal{O})$ are full lattices in $\mathcal{L}(K)$ closed under the Lie bracket such that $[\mathcal{L}_i, \mathcal{L}_j] \subseteq \mathcal{L}_{i+j}$ with equality if $\mathrm{char}(F)$ divides none of the $m_{\alpha\beta}$ where $[X_\alpha, X_\beta] = m_{\alpha\beta} X_{\alpha+\beta}$ and $\alpha, \beta, \alpha + \beta \in \Phi$.

To define a filtration for P let $j = (k+1)q + r$ as above. Then define

$$P_j := \langle \mathcal{X}_{\alpha,q+1}, \mathcal{X}_{\beta,q+2}, \mathcal{X}_{\gamma,q}, \mathcal{X}_{\delta,q+1}, \mathcal{H}_{q+1} \mid \alpha \in \Phi^i, \beta \in \Phi^{-k-1+i} \text{ for } i = 1, \ldots, j-1,$$

$$\gamma \in \Phi^i, \delta \in \Phi^{-k-1+i} \text{ for } i = j, \ldots, k \rangle.$$

To investigate when $P_j = \gamma_j(P)$ we need two lemmas.

(V.2) Lemma. *Let* $G = \varprojlim G/G_i$ *be a pro-p-group with* $G = G_1 \geq G_2 \geq \cdots$ *a central series of subgroups of finite index in* G. *If* $[G_i, G]G_{i+2} = G_{i+1}$ *for all* $i \geq 1$ *then* $\gamma_i(G) = G_i$, *where* $\gamma_i(G)$ *is understood to be the closure of the* i^{th} *term of the lower central series.*

Proof. One has $G_{i+1} = [G_i, G]G_{i+2} = [G_i, G][G_{i+1}, G]G_{i+3} = [G_i, G]G_{i+3}$. An obvious induction yields $G_{i+1} = [G_i, G]G_{i+n}$ for any $n \geq 2$. Hence

$$G_{i+1} = \bigcap_{n \geq 1} [G_i, G]G_{i+n} = \overline{[G_i, G]}.$$

q.e.d.

(V.3) Lemma. *The following commutator relations hold in P. (Convention $gh = hg[g,h]$.)*

(i) Let $\alpha, \beta \in \Phi$ linearly independent. Then

$$[x_\alpha(u), x_\beta(t)] = \prod_{\substack{i,j \in \mathbb{N} \\ i\alpha + j\beta \in \Phi}} x_{i\alpha+j\beta}(c_{ij\alpha\beta} u^i t^j)$$

with $c_{ij\alpha\beta} \in \mathbb{Z}$, in particular $c_{ij\alpha\beta} = m_{\alpha\beta}$ where $[X_\alpha, X_\beta] = m_{\alpha\beta} X_{\alpha+\beta}$.

(ii)

$$[x_\alpha(s), x_{-\alpha}(t)] = x_\alpha\left(\frac{s^2 t}{1-st}\right) h_\alpha(\frac{1}{1-st}) x_{-\alpha}\left(\frac{-st^2}{1-st}\right)$$

where $h_\alpha(u)\bar{X}_\beta = u^{\langle\beta,\alpha\rangle}\bar{X}_\beta$ and $h_\alpha(u)\bar{H}_\beta = \bar{H}_\beta$.

(iii) $[h(\chi_1), h(\chi_2)] = 1$ for $\chi_1, \chi_2 \in \mathrm{Hom}(\mathbb{Z}\Phi, \mathcal{O}_1^*)$.

(iv) $[h_\alpha(u), x_\beta(s)] = x_\beta((1 - u^{-\langle\beta,\alpha\rangle})s)$ for $\alpha, \beta \in \Phi, u \in \mathcal{O}^*, s \in \mathcal{O}$.

Proof. ii) can be checked by calculations carried out in SL_2.
The other equations follow from relations given by [Ste 68]. q.e.d.

As an immediate consequence of (V.3) one sees that $[P, P_i] \subseteq P_{i+1}$.

As the proof of (V.4) will show, there are three possible obstacles for P_i to being equal to $\gamma_i(P)$:

(i) $p \mid m_{\alpha,\beta}$ for some roots α and β such that $\alpha + \beta$ is a root, where $[X_\alpha, X_\beta] = m_{\alpha\beta} X_{\alpha+\beta}$,

(ii) $p \mid |Q : \mathbb{Z}\Phi|$ where Q is the weight lattice of Φ,

(iii) there is a root $\beta \in \Phi^1 \cup \Phi^{-k}$ for which there is no $\alpha \in \Phi$ with $\gcd(\langle\beta,\alpha\rangle, p) = 1$.

Remark.
Obstacle (i) occurs only for $(p, \Phi) \in \{(2, B_n), (2, C_n), (2, F_4), (2, G_2), (3, G_2) | n \geq 2\}$
Obstacle (ii) occurs only for $(p, \Phi) \in \{(2, B_n), (2, C_n), (2, D_n), (2, E_7), (3, E_6) | n \geq 2\}$ or $p | (n+1)$ and Φ is of type A_n.
Obstacle (iii) occurs only for $(p, \Phi) \in \{(2, A_1), (2, B_2), (2, C_n) | n \geq 2\}$.
If obstacle (ii) does not occur also (iii) does not occur. On the other hand, if one extends the group by elements of $Hom(\mathbb{Z}\Phi, \mathcal{O}_1^*)$ (i.e. rational automorphisms) in the case of obstacle (ii) one possibly also avoids obstacle (iii).

(V.4) Theorem. *Let $d = |Q : \mathbb{Z}\Phi|$ as above. For $P = P(\Phi, \mathcal{O})$ one has $P_i = \gamma_i(P)$ for all $i \in \mathbb{N}$, unless $p \mid d$, or $p = 3$ and $\Phi = G_2$, or $p = 2$ and $\Phi \in \{G_2, F_4, B_n, C_n | n \geq 2\}$.*

Proof. Clearly, if one of (i), (ii), or (iii) holds one has one of the exceptions listed in the theorem. Therefore we have to prove: if none of (i), (ii), and (iii) is satisfied,

then $[P, P_i]P_{i+2} = P_{i+1}$ for $i = 1, 2, \ldots$ and the result follows from (V.2). We use induction on i. The three cases considered yield the base and the induction step and the result follows by Lemma (V.2).
Case 1: $i \not\equiv 0, k \bmod k+1$.
In this case (V.3)(i) yields $[P, P_i]P_{i+2} = P_{i+1}$ since condition (i) above does not hold.
Case 2: $i \equiv k \pmod{k+1}$ say $i = q(k+1) + k$.
In this case (V.3)(ii) shows that $h_\alpha(x)$ lies in $[P, P_i]P_{i+2}$ for all $\alpha \in \Phi$, since the x_α- and $x_{-\alpha}$-terms lie in P_{i+2}. Therefore we have to prove
$\langle h_\alpha(x) | \alpha \in \Phi, x \in \mathcal{O}^*_{q+1} \rangle = \mathcal{H}_{q+1}$. The definition of the $h_\alpha(x)$ shows $h_\alpha(x) = \prod h(\chi_\alpha)^{\langle\alpha,\beta\rangle}$ with $\chi_\alpha \in Hom(\mathbb{Z}\Phi, \mathcal{O}^*_{q+1})$ defined by $\chi_\alpha(\beta) = 1$ for $\beta \in \Delta$, $\beta \neq \alpha$ and $\chi_\alpha(\alpha) = x$. Since the $h(\chi_\alpha)$ with $\alpha \in \Delta$ obviously generate $\mathcal{H}_{q+1}$, this holds for the $h_\alpha(x)$ with $x \in \mathcal{O}^*_{q+1}$ as well, if $p \not| \det(\langle\beta, \alpha\rangle)_{\alpha,\beta\in\Delta} = |Q : \mathbb{Z}\Phi|$, which is the negation of (ii) above.
Case 3: $i \equiv 0 \pmod{k+1}$.
Here relation (V.3)(iv) has to be applied. Since condition (iii) above is not satisfied one gets $[P, P_i]P_{i+2} = P_{i+1}$ in this case. q.e.d.

One should remark - especially important for the case $\Phi = A_n$ - that Theorem (V.4) remains valid if one modifies $P = P(\Phi, \mathcal{O})$, by working with the Sylow pro-p-subgroup of the Chevalley group rather than with the Sylow pro-p-subgroup of $\mathrm{Aut}(\mathcal{L}(K))$. (Of course the $\mathcal{H}_i$ have to be modified accordingly. Note, however that the Sylow subgroup is no longer just infinite since it gets a finite centre and even after factoring out this centre it is not a maximal $\tilde{p}$-group in case of $char(K) = 0$.) Then condition (ii) becomes redundant and one gets $P_i = \gamma_i(P)$ except for the cases
$(p, \Phi) \in \{(2, A_1), (2, B_n), (2, C_n), (2, F_4), (2, G_2), (3, G_2) \mid n \geq 2\}$.

VI Some classical groups

Most of the classical groups can be studied within the context of simple algebras with involution (cf. [Wei 61], [Kne 69]), which at the same time provide the context for the Cayley maps (cf. [Wey 46]) connecting the unitary group with its Lie algebra if the characteristic is not 2. In the present situation also orders in the algebra which are invariant under the involution have to be considered. A general reference on orders is [Rei 75].
Apart from the general notation the first section can be skipped if one is only interested in the results, in particular in the orders of the factor groups of the lower central series of the classical groups. Example (VI.18) explains how to read Table (VI.17) which contains information about the generators of a Sylow pro-p-subgroup of the classical groups, about the lower central series, about the involution $^\circ$ by which the groups are defined and about a minimal $^\circ$-invariant order. The patterns of the lower central series are summarised in Table (VI.19) and can be understood without anything of the previous sections.

a) Basic structures: orders and Cayley maps

Notation and assumption:

- *K a local field with ring of integers $\mathcal{O}$ and a prime element Π.*
- *$p \neq 2$ the characteristic of $(\mathcal{O}/\Pi\mathcal{O})$.*
- *$\bar{\ } \in Aut(K)$ with $\bar{\ }^2 = id_K$.*
- *k the fixed field of $\bar{\ }$, $o = k \cap \mathcal{O}$, π is a prime element of k.*

Three situations are possible:

(i) $\bar{\ } = id$, then $K = k, \mathcal{O} = o$ and let $\Pi = \pi$.

(ii) $\bar{\ } \neq id$, K/k unramified. In this case $\mathcal{O} = o \oplus oq$ with $q^2 \in o - \pi o$ and let $\Pi = \pi$.

(iii) $\bar{\ } \neq id$, K/k ramified. In this case $\mathcal{O} = o \oplus o\Pi$, assume $\Pi^2 = \pi$.

Note, in (ii) there is one and in (iii) there are two isomorphism types for K for a given k.

- *$(A, ^\circ)$ a central simple K-algebra with involution $^\circ$ such that $^\circ$ induces $\bar{\ }$ on K i.e. $^\circ$ is a k-anti-automorphism on A fixing k and its square equals the identity. (Remember, if $\bar{\ }$ is the identity on K then $^\circ$ is called to be of first kind otherwise $^\circ$ is called to be of second kind.)*
- *V a simple A-module.*
- *$A^+ := \{x \in A | x^\circ = x\}$.*
  $A^- := \{x \in A | x^\circ = -x\}$.

Note, A^- is a k-Lie algebra with $[x,y] := xy - yx$ for all $x, y \in A$.

- $U := U(A,^\circ) := U(A) := \{x \in A | xx^\circ = 1_A\}$ *unitary group of* $(A,^\circ)$.
- $SU := SU(A,^\circ) := SU(A) := \{x \in U | \mathrm{nr}(x) = 1_k\}$ *where* nr *is the reduced norm of* A *over* k.
- $SA := \{x \in A | \mathrm{tr}(x) = 0\}$ *where* tr *is the reduced trace of* A *over* k.
- $SA^- := \{x \in A^- | \mathrm{tr}(x) = 0\}$ *where* tr *is the reduced trace of* A *over* k.
- $X_{gen} := \{x \in X | -1$ *is not an eigenvalue of* $x\}$ *for any subset* X *of* A.
- $c_{PL} : U_{gen} \to A^-_{gen} : x \mapsto (1-x)(1+x)^{-1}$, $c_{LP} : A^-_{gen} \to U_{gen} : x \mapsto (1-x)(1+x)^{-1}$ *are the Cayley maps, which are easily seen to be inverse to each other and compatible with conjugation by elements of* U.

The aim of this chapter is to compute the lower central series of the maximal pro-p-subgroup of SU. To this end these groups have to be described rather explicitly in terms of certain Lie lattices in A^- which turn out to be of the form Λ^- for certain carefully chosen $^\circ$-invariant o-orders Λ in A. The powers of the radical of these orders turn out to be closely connected to the lower central series of the corresponding Sylow pro-p-subgroup of $U(A)$. The link between the Lie lattices and the pro-p-groups is provided by the Cayley maps. The $^\circ$-invariant orders are rather close to the o-span of the groups. There is an unpleasant but not so crucial side-effect in the case where the involution is of second kind and therefore the centre $Z(U)$ is not finite. We are really interested in the maximal pro-p-subgroups of $PU := U/Z(U)$ with $Z(U) = \{x \in K = Z(A) | x\bar{x} = 1\}$ or PU extended by outer p-automorphisms. However, one shall investigate the maximal pro-p-subgroups of SU, which often amounts to the same. By the general theory, cf. Lemma (III.16), SU has maximal pro-p-subgroups and they are all conjugate under SU.

It is worth while to set up correspondences between

- $\mathcal{P} = \{P | P$ *is an open pro-p-subgroup of* $U\}$,
- $\mathcal{D} = \{\Lambda | \Lambda$ *is an* o*-order in* A *with* $\Lambda^\circ = \Lambda\}$, *and*
- $\mathcal{L} = \{L | L$ *is a full* o*-Lie-lattice in* A^- *with* $(\ldots(ML)L\ldots) \subsetneq \pi M$ *for some* o*-lattice* M *in* $V\}$.

Note, for $L \in \mathcal{L}$ every element of L is ad-pro-nilpotent.

Clearly, one has the following maps:

$$\mathcal{D} \to \mathcal{P} : \Lambda \mapsto (1 + (\mathrm{rad}\Lambda)) \cap U =: U(\Lambda),$$

$$\mathcal{D} \to \mathcal{L} : \Lambda \mapsto (\mathrm{rad}\Lambda)^- := \{x \in \mathrm{rad}\Lambda | x^\circ = -x\}.$$

Neither map is surjective. There is an obvious modification of the maps, if one works with SU and SA^-.

Next there is an obvious map

$$\mathcal{L} \to \mathcal{D} : L \mapsto \langle L \rangle_{o\text{-order}} = Ord(L).$$

For $L \in \mathcal{L}$ let $\bar{L} = \mathrm{rad}(Ord(L))^-$. Clearly $L \subseteq \bar{L} \in \mathcal{L}$ and $\bar{\bar{L}} = \bar{L}$.

Next one defines a subclass $\tilde{\mathcal{L}}$ of $\mathcal{L}$ for which the Cayley map is well defined. Define $\tilde{\mathcal{L}} = \{L \in \mathcal{L} | L \text{ satisfies } (*)\}$ where $(*)$ is the following condition.

$(*)$ For all $x, y \in L$ and $p(X,Y) \in \mathbb{Z}\langle X, Y\rangle$ (= free associative algebra on X, Y) homogenous of degree $k \geq 1$, then $p(x,y) \in L$ provided $p(x,y)^\circ = -p(x,y)$.

Clearly $L \in \mathcal{L}$ implies $\bar{L} \in \tilde{\mathcal{L}}$. More generally one obviously has

<u>(VI.1) Remark.</u> *If $\Lambda \in \mathcal{D}$ is an o-order with $\Lambda^\circ = \Lambda$, and $I \lhd \Lambda$ with $I \subseteq \mathrm{rad}\Lambda$ and $I^\circ = I$ then $I^- \in \tilde{\mathcal{L}}$.*

Now use the second Cayley map to construct a map

$$C_{LP} : \tilde{\mathcal{L}} \to \mathcal{P} : L \mapsto Lc_{LP} = \{(1-x)(1+x)^{-1} | x \in L\}$$

<u>(VI.2) Lemma.</u> *The map C_{LP} is well defined.*

Proof. Clearly $L \in \mathcal{L} \Rightarrow L \subseteq A^-_{gen}$ since all eigenvalues of $x \in L$ are congruent 0 (mod πo). Therefore xc_{LP} is well defined and $xc_{LP} \in U$. Clearly $(-x)c_{LP} = (xc_{LP})^{-1}$. Let M be a full o-lattice in V with $(\ldots(ML)\ldots)L \subseteq M\pi$. Then clearly xc_{LP} centralises an (full) o/π-flag of $M/\pi M$. Therefore LC_{LP} is contained in an open pro-p-subgroup of U. To prove that LC_{LP} is open, note that $c_{LP} : U_{gen} \to A^-_{gen}$ is a homeomorphism and therefore maps the open subset L of A^- onto the open set LC_{LP}. That LC_{LP} is a group requires $L \in \tilde{\mathcal{L}}$ and follows now from the next lemma. q.e.d.

<u>(VI.3) Lemma.</u> *Let $L \in \tilde{\mathcal{L}}$. Then for any $x, y \in L$ there exists a $z \in L$ with $xc_{LP}yc_{LP} = zc_{LP}$. More precisely, there exist $p_k(X,Y) \in \mathbb{Z}\langle X,Y\rangle$ homogeneous of degree k for $k = 1, 2, \ldots$, such that $z = \sum_{k=1}^\infty p_k(x,y)$.*
E.g. $p_1 = X + Y, p_2 = -XY + YX, p_3 = -XYX - YXY, p_4 = (XY)^2 - (YX)^2, p_5 = (XY)^2X + (YX)^2Y$ etc..
(The final formula for the p_k is established in [Sou 96]*.)*

Proof. $xc_{LP} = (1-x)\sum_{i=0}^\infty (-x)^i = 1 + 2\sum_{i=1}^\infty (-x)^i := 1 + 2\tilde{x}$
and similarly $yc_{LP} = (1-y)\sum_{j=0}^\infty (-y)^j = 1 + 2\tilde{y}$. Hence
$xc_{LP}yc_{LP} = 1 + 2(\tilde{x} + \tilde{y} + 2\tilde{x}\tilde{y})$. With $\tilde{z} := \tilde{x} + \tilde{y} + 2\tilde{x}\tilde{y}$ one gets

$$(xc_{LP}yc_{LP})c_{PL} = (1 - (1 + 2\tilde{z}))(1 + (1 + 2\tilde{z}))^{-1} = -\tilde{z}\sum_{k=0}^{\infty}(-\tilde{z})^k =: z.$$

Clearly everything converges, $z \in L$ because of $(*)$, and $z = \sum_{k=1}^n z_k$ with $z_k^\circ = -z_k$ a Z-combination of products of x and y with k factors.

Repeating the calculation in a suitable completion of $\mathbb{Z}\langle X, Y\rangle$ and collecting the monomials of the same degree together, one gets the desired formula for $z_1, z_2, \ldots$. q.e.d.

Using the composite $\mathcal{D} \to \bar{\mathcal{L}} \to \mathcal{P}$ one can exhibit a central series of the groups constructed this way, also this composite is equal to the map $\mathcal{D} \to \mathcal{P}$ introduced above.

(VI.4) Definition. *Let* $\Lambda \in \mathcal{D}$ *be an* $^\circ$*-invariant* o*-Order in* A. *Define* $J_i^- = J_i^-(\Lambda) := (\mathrm{rad}\Lambda)^i \cap A^-$ *and* $U_i(\Lambda) := (1 + (\mathrm{rad}\Lambda)^i) \cap U$.

The two filtrations of $(\mathrm{rad}\Lambda)^-$ and $U(\Lambda)$ have the following properties.

(VI.5) Proposition.

(i) $U(\Lambda) = U_1(\Lambda) \geq U_2(\Lambda) \geq \cdots$ *with* $U_i(\Lambda) \trianglelefteq U(\Lambda)$ *open,* $U_i(\Lambda)/U_{i+1}(\Lambda)$ *elementary abelian p-groups,* $[U_i(\Lambda), U_j(\Lambda)] \subseteq U_{i+j}(\Lambda)$ *and* $\cap_{i\in\mathbb{N}} U_i(\Lambda) = \{1\}$

(ii) $(\mathrm{rad}\Lambda)^- = J_1^- \geq J_2^- \geq \cdots$ *with* J_i^-/J_{i+1}^- *finite* $o/\pi o$ *- modules such that* $[J_i^-, J_j^-] \subseteq J_{i+j}^-$ *for* $i, j \in \mathbb{N}$, *and* $\cap_{i\in\mathbb{N}} J_i^- = \{0\}$. *Moreover* $J_i^-/J_{i+1}^- \cong (J_i/J_{i+1})^- :=$ (-1)*-eigenspace of* $^\circ$ *acting on* J_i/J_{i+1} *with* $J_i = (\mathrm{rad}\Lambda)^i$.

(iii) $J_i^- \in \bar{\mathcal{L}}$ *and* $J_i^- C_{LP} = U_i(\Lambda)$ *for* $i \in \mathbb{N}$.

(iv) The second Cayley map c_{LP} *induces an isomorphism* $J_i^-/J_{i+1}^- \to U_i(\Lambda)/U_{i+1}(\Lambda)$ *for each* $i \in \mathbb{N}$.

Proof. (ii) J_i/J_{i+1} is a finite $o/\pi o$ -module; therefore J_i^-/J_{i+1}^- is a finite $o/\pi o$-module, and since $2 \neq p = \mathrm{char}(o/\pi)$ one gets immediately $(J_i/J_{i+1})^- \cong J_i^-/J_{i+1}^-$. Since $J_iJ_j = J_{i+j}$ one gets $[J_i^-, J_j^-] \subseteq J_{i+j}^-$. Finally $\cap_{i\in\mathbb{N}} J_i^- \subseteq \cap J_i = \{0\}$.

(iii) By (VI.1) $J_i^- \in \bar{\mathcal{L}}$ and hence $J_i^- C_{LP} \in \mathcal{P}$ is well defined.
Since $x c_{LP} = 1 - 2x + 2x^2 - 2x^3 \pm \cdots$ for $x \in J_i^-$, one has $J_i^- C_{LP} \leq U_i(\Lambda)$. Since $p \neq 2$, $U_i(\Lambda) \subseteq U_{gen}$, hence $U_i(\Lambda) c_{PL}$ is defined. Let $g \in U_i(\Lambda)$. With $y = (1/2)(g-1) \in J_i$ one gets

$$g c_{PL} = (1-g)(1+g)^{-1} = -2y\frac{1}{2}(1+y)^{-1} = -y(1 - y + y^2 - \cdots) \in J_i^-.$$

Since the Cayley maps c_{PL} and c_{LP} are inverse to each other, one has $U_i(\Lambda) = J_i^- C_{LP}$.

(i) By (VI.2) and (iii) $U_i(\Lambda)$ is open in U, as also the definition shows directly. It remains to prove $[U_i(\Lambda), U_j(\Lambda)] \subseteq U_{i+j}(\Lambda)$. For $g \in U_i(\Lambda)$ and $h \in U_j(\Lambda)$ write $g = 1 + x$ and $h = 1 + y$ with $x \in J_i$ and $y \in J_j$. Hence $g^{-1} = 1 - x + x^2 - \cdots$, $h^{-1} = 1 - y + y^2 - y^3 + \cdots$ yields the following expansion for the group commutator:

$$[g, h] = g^{-1}h^{-1}gh = 1 + xy - yx + z$$

with $z \in J_{i+j+1}$. Hence $[g, h] \in U_{i+j}(\Lambda)$, more precisely $[g, h] \equiv 1 + xy - yx \bmod 1 + J_{i+j+1}$.

(iv) This now follows immediately from the expansion of C_{LP} in the proof of (iii). q.e.d.

It will turn out that for suitable choices for Λ, cf. (VI.5) even gives the lower central series of the maximal pro-p-subgroup of U or SU. But first continue with the general discussion. Having constructed groups out of orders and Lie rings, now one goes in the other direction starting with groups.

$$\mathcal{P} \to \mathcal{D} : P \mapsto oP := \langle P \rangle_o$$

$=$ o-span of the elements of P in A is clearly well defined. It gives rise to a saturation process: For $P \in \mathcal{P}$ let $\bar{P} = (1+\mathrm{rad}(oP)) \cap U$. Clearly, $P \subseteq \bar{P} \in \mathcal{P}$. A straightforward calculation shows $\bar{\bar{P}} = \bar{P}$. Define $\bar{\mathcal{P}} = \{\bar{P} | P \in \mathcal{P}\}$. The final map one has to discuss is induced by the first Cayley map c_{PL}.

$$C_{PL} : \bar{\mathcal{P}} \to \bar{\mathcal{L}} : P \mapsto PC_{PL} = \{gc_{PL} | g \in P\}$$

(VI.6) Remark. *The map C_{PL} is well defined.*

Proof. As observed earlier, $p \neq 2$ implies that $P \in \mathcal{P}$ lies in U_{gen}. Obviously $PC_{PL} \subseteq (\mathrm{rad}(oP))^-$, cf. proof of (VI.5)(iii). Let $x \in (\mathrm{rad}(oP))^-$. Then $xc_{LP} = (1-x)(1+x)^{-1} \in (1+\mathrm{rad}(oP)) \cap U$, since both $1+x, 1-x \in 1+\mathrm{rad}(oP) \subseteq (oP)^*$. Hence $P \subseteq (\mathrm{rad}(oP)^-)C_{PL} \subseteq \bar{P}$, but $P = \bar{P}$. Hence $PC_{PL} = (\mathrm{rad}(oP))^- \in \bar{\mathcal{L}}$. q.e.d.

Summarising one has :

(VI.7) Proposition.

(i) The maps

$$C_{LP} : \bar{\mathcal{L}} \to \bar{\mathcal{P}} \text{ and } C_{PL} : \bar{\mathcal{P}} \to \bar{\mathcal{L}}$$

induced by the Cayley maps are inverse to each other.

(ii) C_{LP} factors $\bar{\mathcal{L}} \to \mathcal{D} \to \bar{\mathcal{P}} : L \mapsto ord(L) \mapsto (1+\mathrm{rad}(ord(L))) \cap U$

(iii) C_{PL} factors $\bar{\mathcal{P}} \to \mathcal{D} \to \bar{\mathcal{L}} : P \mapsto oP \mapsto (\mathrm{rad}(oP))^-$.

In the case of involutions of the second kind, i.e. $k \neq K$, one has a version for SU rather than U. We leave the necessary changes to the reader. Note, however, the Cayley maps do not necessarily map trace 0 elements to norm 1 elements and vice versa. However, if every eigenvalue a of $x \in A^-$ occurs with the same multiplicity as $-a$ then $\mathrm{nr}(ac_{LP}) = 1$.

Our analysis, cf. in particular the proof of Lemma (VI.2), suggests that one may construct the maximal pro-p-subgroups P of U from certain $^\circ$-invariant hereditary orders Λ. Note, for an hereditary o-order Λ in A the Λ-lattices M_i in V with $i \in \mathbb{Z}$, which are ordered by inclusion the numbering can be chosen such that $M_{i+1} \subseteq M_i$ and $M_{i+1} = M_i\mathrm{rad}\Lambda$. Moreover $\mathrm{End}_\Lambda(M_i)$ is the maximal $\mathcal{O}$-order in $\mathrm{End}_A(V)$ and therefore Λ contains $\mathcal{O}$. Also $\Lambda = \{x \in A | M_i x \subseteq M_i \text{ for all } i \in \mathbb{Z}\}$ and $\mathrm{rad}\Lambda = \{x \in A | M_i x \subseteq M_{i+1} \text{ for all } i \in \mathbb{Z}\}$. There is an $\alpha \in \mathbb{N}$ with $M_i\Pi = M_{i+\alpha}$ for all $i \in \mathbb{Z}$, which at the same time is the number of simple Wedderburn components of $\Lambda/\mathrm{rad}\Lambda$, cf. [Rei 75] for details. Note, if Λ_1, Λ_2 are hereditary $\mathcal{O}$-orders in A, then $\Lambda_1 \subseteq \Lambda_2$ if and only if $\mathrm{rad}\Lambda_1 \supseteq \mathrm{rad}\Lambda_2$.

(VI.8) Corollary. *Let $P \subseteq U$ be a maximal pro-p-subgroup of U. Then the following hold:*

(i) $P \in \bar{\mathcal{P}}$.

(ii) There is an hereditary $\mathcal{O}$-order Λ with $\Lambda^\circ = \Lambda$ and $(\mathrm{rad}\Lambda)^- = (\mathrm{rad}(oP))^-$.

Proof. (i) $\bar{P} = P$ since P is maximal.
(ii) oP is contained in an $^\circ$-invariant hereditary order, cf. [Scha 74]. Among these choose one with $(\mathrm{rad}\Lambda)^-$ maximal. By the maximality of P one gets $(\mathrm{rad}(oP))^- = (\mathrm{rad}\Lambda)^-$. q.e.d.

(VI.9) Definition. *Let P and Λ be as in (VI.8)(ii). We call Λ semi-saturated if $M_i(\mathrm{rad}\Lambda)^- = M_{i+1}$ for all $i \in \mathbb{Z}$, where $\cdots M_i > M_{i+1} \cdots$ are the Λ-lattices in V. We call Λ saturated, if $SA \cap ((\mathrm{rad}\Lambda)^i)^- = ((\mathrm{rad}\Lambda)^-)^{[i]}$, where $((\mathrm{rad}\Lambda)^-)^{[i]}$ is the o-span of the $[\ldots[l_1, l_2], l_3 \ldots, l_i]$ with $l_1, \ldots, l_i \in (\mathrm{rad}\Lambda)^-$ and $i \in \mathbb{N}$.*

Clearly, Λ is saturated if and only if the defining condition is satisfied for $i = 1, \ldots, \alpha e - 1$ where α is the number of components of $\Lambda/\mathrm{rad}\Lambda$ and e is the ramification index of k in K. Also if Λ is saturated, it is semi-saturated. One can view the semi-saturated orders as those $^\circ$-invariant hereditary orders which are maximal with respect to the property $(\mathrm{rad}\Lambda)^- = (\mathrm{rad}(oP))^-$.
As a result of going through all possible K-algebras with involution, it will turn out that P can always be obtained from a semi-saturated hereditary order, which with one exception for $p = 3$ is already saturated and gives a description of the lower central series of P.

We now derive a formula for Lie commutators, adapted to hereditary orders. Assume $A = D^{n \times n}$ where D is a division algebra with maximal order Ω and radical $\mathbf{P} = \mathrm{rad}\Omega$. An hereditary order Λ can be assumed to be of the form

$$\Lambda = \{(a_{ij}) | a_{ij} \in \Omega^{n_i \times n_j} \text{ for } 0 \leq i \leq j \leq \alpha - 1 \text{ and } a_{ij} \in \mathbf{P}^{n_i \times n_j} \text{ for } 0 \leq j < i \leq \alpha - 1\}$$

where $n_0, \ldots, n_{\alpha-1} \in \mathbb{N}$ with $n_0 + \cdots + n_{\alpha-1} = n$. With $n_0, \ldots, n_{\alpha-1}$ fixed, $(m_{ij}) \in (\mathbb{N} \cup \{0\})^{\alpha \times \alpha}$ and $\mathbf{P}^0 := \Omega$ let

$$\Lambda((m_{ij})) = \{(a_{ij}) | a_{ij} \in (\mathbf{P}^{m_{ij}})^{n_i \times n_j} \text{ for } 0 \leq i, j \leq \alpha - 1\}.$$

$$\text{Then } \Lambda = \Lambda\left(\begin{pmatrix} 0 & \cdot & \cdot & \cdot & \cdot & 0 \\ 1 & \cdot & & & & \cdot \\ \cdot & & \cdot & & & \cdot \\ \cdot & & & \cdot & & \cdot \\ \cdot & & & & \cdot & \cdot \\ 1 & \cdot & \cdot & \cdot & 1 & 0 \end{pmatrix}\right),\ \mathrm{rad}\Lambda = \Lambda\left(\begin{pmatrix} 1 & 0 & \cdot & \cdot & \cdot & 0 \\ \cdot & \cdot & & & & \cdot \\ \cdot & & \cdot & & & \cdot \\ \cdot & & & \cdot & & \cdot \\ \cdot & & & & \cdot & 0 \\ 1 & \cdot & \cdot & \cdot & \cdot & 1 \end{pmatrix}\right),$$

$$(\mathrm{rad}\Lambda)^2 = \Lambda(\begin{pmatrix} 1 & 1 & 0 & . & . & 0 \\ . & . & . & . & . & . \\ . & & & . & . & . \\ . & & & . & . & 0 \\ 1 & & & & . & 1 \\ 2 & 1 & . & . & . & 1 \end{pmatrix}), \ldots, (\mathrm{rad}\Lambda)^i = \Lambda(E_i),$$

where E_i is obtained from E_{i-1} by shifting all columns one position to the right and taking the last column of E_{i-1} with entries increased by 1 as first column of E_i. Call the matrix (m_{ij}) exponent matrix of the order or ideal.

We introduce some notation for representatives of cosets in $(\mathrm{rad}\Lambda)^i/(\mathrm{rad}\Lambda)^{i+1}$.

(VI.10) Definition. *Let $0 \le i \le \alpha - 1, a_0 \in D^{n_0 \times n_i}, a_1 \in D^{n_1 \times n_{i+1}}, \ldots,$ $a_{\alpha-1} \in D^{n_{\alpha-1} \times n_{i+\alpha-1}}$, where the indices s of n_s are taken modulo α. Then*

$$d_i(a_0, \ldots, a_{\alpha-1}) := \left(\begin{array}{cccccccc} & & & & a_0 & . & . & 0 \\ & & & & 0 & . & & . \\ & 0 & & & . & & . & . \\ & & & & . & & . & 0 \\ & & & & 0 & . & 0 & a_{\alpha-i-1} \\ a_{\alpha-i} & 0 & . & 0 & & & & \\ 0 & . & & . & & & & \\ . & & . & . & & 0 & & \\ . & & . & 0 & & & & \\ 0 & . & 0 & a_{\alpha-1} & & & & \end{array}\right) \in D^{n \times n},$$

where a_l is in the l-th block row and in the i-th block column. (In particular $d_0(a_0, \ldots, a_{\alpha-1}) = diag(a_0, \ldots, a_{\alpha-1})$.)

If the entries of the a_i lie in the appropriate powers of $\mathbf{P}$ (as indicated by the exponent matrix E_i), then $d_{i'}(a_0, \ldots, a_{\alpha-1}) \in (\mathrm{rad}\Lambda)^i$, where $i' \equiv i \bmod \alpha$. An elementary calculation yields the following formula for Lie brackets.

(VI.11) Lemma. *For $0 \le i, j \le \alpha - 1$ and $a_0 \in D^{n_0 \times n_i}, a_1 \in D^{n_1 \times n_{i+1}}, \ldots, a_{\alpha-1} \in D^{n_{\alpha-1} \times n_{\alpha+i}}, b_0 \in D^{n_0 \times n_j}, \ldots, b_{\alpha-1} \in D^{n_{\alpha-1} \times n_{\alpha+j}}$, then*

$$[d_i(a_0, \ldots, a_{\alpha-1}), d_j(b_0, \ldots, b_{\alpha-1})] = d_{i+j}(c_0, \ldots, c_{\alpha-1})$$

with $i + j$ taken mod α to lie between 0 and $\alpha - 1$, and

$$c_l = a_l b_{l+i} - b_l a_{l+j}$$

for $0 \le l \le \alpha - 1$, with all indices taken mod α.

Clearly, with this formula it becomes easy to check whether an hereditary °-invariant order Λ is saturated. We now translate Lie commutators into group commutators.

(VI.12) Lemma. *Let $L \in \tilde{\mathcal{C}}$. Then there are $p_i(X, Y) \in \mathbb{Z}\langle X, Y\rangle$ homogenous of degree i for $i \ge 2$ such that for any $x, y \in L$ the group commutator is given by*

$$[xc_{LP}, yc_{LP}] = 1 + p_2(x, y) + p_3(x, y) + \cdots$$

The first two p_i *are* $p_2(X,Y) = 4(XY - YX) = 4[X,Y]$,
$p_3(X,Y) = 4(X^2Y - 2XYX - XY^2 + YX^2 + 2YXY - Y^2X) = -4[[X,Y],X+Y]$.
(The Lie expression of the p_i *is established in* [Sou 96].*)*

Proof. $[xc_{LP}, yc_{LP}] = (xc_{LP})^{-1}(yc_{LP})^{-1}xc_{LP}yc_{LP} = (-x)c_{LP}(-y)c_{LP}xc_{LP}yc_{LP} = (1+2x+2x^2+2x^3+\cdots)(1+2y+2y^2+2y^3+\cdots)(1-2x+2x^2-2x^3+\cdots)(1-2y+2y^2-2y^3+\cdots) = 1+4[x,y]+z_3+z_4\cdots$ and the claim follows similarly as in the proof of (VI.5). q.e.d.

This was the last preparation for the computation of the lower central series of a maximal pro-p-subgroup of SU.

(VI.13) Proposition. *Let* $\Lambda = \Lambda^\circ$ *be a saturated hereditary* $\mathcal{O}$*-order in* A*. Then the maximal pro-p-subgroup* $P = SU \cap (\mathrm{rad}\Lambda)^- C_{LP}$ *of* SU *has the lower central series given by* $P_1 = P, P_i = SU \cap (1 + (\mathrm{rad}\Lambda)^i) = SU \cap ((\mathrm{rad}\Lambda)^i)^- C_{LP}$.

Proof. This follows immediately from (VI.5), the definition of saturated orders, (VI.12), and (V.2). q.e.d.

Before going through the various cases, some general comments on the description of the involution $^\circ$ of $A = D^{n\times n}$ leaving an hereditary order Λ invariant are in place.

(VI.14) Remark. *Let* $\Lambda = \Lambda^\circ$ *be an hereditary* $\mathcal{O}$*-order in* A*. Then the permutation induced by the involution* $^\circ$ *on the set of the Wedderburn components of* $\Lambda/\mathrm{rad}\Lambda$ *has at most 2 fixed points.*

Proof. Let $\Lambda/\mathrm{rad}\Lambda = X_0 \oplus \cdots \oplus X_{\alpha-1}$ be the decomposition of $\Lambda/\mathrm{rad}\Lambda$ into simple algebras and write $X_i^\circ = X_{i\iota}$. The numbering can be chosen in such a way that $\mathrm{rad}\Lambda/(\mathrm{rad}\Lambda)^2 = Y_{0,1} \oplus Y_{1,2} \oplus \cdots \oplus Y_{\alpha-1,0}$ where $Y_{i,i+1}$ is a X_i-X_{i+1}-bimodule. Applying the involution $^\circ$ yields $Y_{i,i+1}^\circ = Y_{(i+1)\iota,i\iota}$, and this is a $X_{(i+1)\iota}$-$X_{i\iota}$-bimodule. Assume $i_0\iota = i_0$, then $(i_0+j)\iota = i_0 - j$ (all indices taken modulo α). The claim follows. q.e.d.

The decomposition of $\Lambda/\mathrm{rad}\Lambda$ enables one to come up with a rather refined notion of the $o/\pi o$-dimension of $((\mathrm{rad}\Lambda)^i/(\mathrm{rad}\Lambda)^{i+1})^-$ as follows:

$$(\mathrm{rad}\Lambda)^i/(\mathrm{rad}\Lambda)^{i+1} = Y_{0,i}^{(i)} \oplus Y_{1,i+1}^{(i)} \oplus \cdots \oplus Y_{\alpha-1,\alpha+i-1}^{(i)}$$

with the convention that all indices are taken modulo α, where $Y_{j,i+j}^{(i)}$ is a X_j-X_{i+j}-bimodule. Clearly ${Y_{j,i+j}^{(i)}}^\circ = Y_{(i+j)\iota,j\iota}^{(i)}$ in the notation of the proof of the last remark.

(VI.15) Definition. *For* $(r-1)\alpha < i \leq r\alpha$, $0 \leq s,t < \alpha$, $s+i \equiv t \bmod \alpha$ *let*

$$b_{s,t}^{(r)} := \begin{cases} \dim_{o/\pi o}(Y_{s,t}^{(i)})^- & \text{if } (s,t) = (t\iota, s\iota) \\ \dim_{o/\pi o}(Y_{s,t}^{(i)}) & \text{if } (s,t) \textit{ is lexicographically earlier than } (t\iota, s\iota) \\ - & \textit{otherwise.} \end{cases}$$

$B_r = (b_{s,t}^{(r)})_{0\leq s,t<\alpha}$ *is called the* r*-th dimension matrix of* $\mathrm{rad}(\Lambda)^-$ *or of* $(\mathrm{rad}\Lambda)^- C_{LP}$.

One clearly has

(VI.16) Remark.

(i) $B_r = B_{r+e}$ *where e is the ramification index of D over K.*

(ii) *For* $i \in \mathbb{N}$ *let* $b_i = \sum_{t-s\equiv i} b_{s,t}^{(r)}$ *where* $s,t \in \{0,1,\dots,\alpha-1\}$, *r defined by* $\alpha(r-1) \leq i < \alpha r$ *with the convention that all indices are to be taken modulo* α. *Then* $\dim_{o/\pi o}(\mathrm{rad}\Lambda)^i/(\mathrm{rad}\Lambda)^{i+1} = b_i$.

So in particular one can read off from the dimension matrices the orders of the sections of the lower central series of $P = SU \cap (\mathrm{rad}\Lambda)^- C_{LP}$, in the case the group is saturated. Note, however, if $\alpha \mid i$ this order might not be p^{fb_i} but $p^{f(b_i-1)}$ where f =degree of $o/\pi o$ over $\mathbb{F}_p$, depending on the difference between Λ^- and $SA \cap \Lambda^-$. If one has to take $b_{\alpha r} - 1$ instead of $b_{\alpha r}$, this is indicated by a • as an exponent to B_r. The concrete way in which the involution $^\circ : A = D^{n\times n} \to A$ is given, is by

$$(a_{ij})_{1\leq i,j\leq n} \mapsto F(\bar{a}_{ij})^{tr} F^{-1}$$

for a suitable $F \in A^*$, which can be thought of as a Gram matrix for some ε-Hermitian form on V with $\varepsilon = \pm 1$. In case D is a K-division algebra it is a non-split quaternion algebra Q and $^-$ is the standard involution. More precisely

- Q has a unique maximal $\mathcal{O}$-order $\Omega = \langle 1, q, \pi_Q, q\pi_Q\rangle_{\mathcal{O}}$
- with $q^2 \in K = k$ a non-square unit, $\pi_Q^2 = \pi$, $q\pi_Q q^{-1} = -\pi_Q$.
- $^-$ maps $1, q, \pi_Q$ and $q\pi_Q$ to $1, -q, -\pi_Q$ and $-q\pi_Q$ respectively.

The special shape of F in the tables below is chosen in such a way that the components of $\Lambda/\mathrm{rad}\Lambda$ and the permutations induced by $^\circ$ on these components, cf. (VI.14), becomes evident.

b) Tables: Involutions, orders and lower central series

In the subsequent table one goes through the various simple algebras with involution (for characteristic 0 the list is complete) and the following information is given:

(0) Dynkin diagram of the unitary group $SU(A,^\circ)$ as defined in [Sat 71] p. 119.

(1) Name of the unitary group, information on the division algebra D, (so that $A = D^{n\times n}$ is described, Q denotes a non-split quaternion algebra).

(2) Matrix F (so that together with (1) $(A,^\circ)$ is described),

(3) α = number of Wedderburn components of $\Lambda/\mathrm{rad}\Lambda$ for the hereditary order Λ. In case $\alpha < n$ those terms of the sequence $(n_0, \dots, n_{\alpha-1})$ which consists of degrees of the components of $\Lambda/\mathrm{rad}\Lambda$ over their centres are given, which are not equal to 1. Then Λ is always of the form described before (VI.10).

(4) The dimension matrices $B = B_1$ or B_1 and B_2 respectively in case the ramification index of D over K is 1 or 2. In case the involution is of the second kind one of the B_i's gets a • as explained after (VI.16).

Abbreviations:

$$\begin{pmatrix} A_{11} & \dots & A_{1l} \\ \vdots & & \vdots \\ A_{l1} & \dots & A_{ll} \end{pmatrix}_{m_1,\dots,m_l}$$

denotes a matrix consisting of submatrices A_{ij} which have m_i rows and m_j columns. Such a submatrix is denoted by a, if all its entries are equal to a. It is denoted by $({}^a\, b\, {}_c)$ if it is square of the form

$$\begin{pmatrix} a & \dots & a & b \\ \vdots & \cdot^{\cdot^{\cdot}} & \cdot^{\cdot^{\cdot}} & c \\ a & \cdot^{\cdot^{\cdot}} & \cdot^{\cdot^{\cdot}} & \vdots \\ b & c & \dots & c \end{pmatrix}$$

which degenerates into (b) if the degree is 1. The possible entries a, b, c can be natural numbers, 0, and $-$, with the meaning explained above for the dimension matrices B_i and elements of D for the Gram matrices F.

(5) The pattern of the lower central series is encoded in the following way. The number of copies of cyclic groups C_p divided by the inertia degree f of k is given for each factor group γ_i/γ_{i+1}. A sequence of numbers enclosed by brackets and raised to some power x means the repetition of this sequence x-times. The whole sequence is overlined to indicate that it repeats itself. E.g. $\overline{n^{n-1}, n-1}$ means:
$\gamma_{1+kn}/\gamma_{2+kn} \cong C_p^{fn}, \dots, \gamma_{n-1+kn}/\gamma_{n+kn} \cong C_p^{fn}, \gamma_{n+kn}/\gamma_{n+1+kn} \cong C_p^{f(n-1)}$ for $k \in \mathbb{N} \cup \{0\}$.

One finds an example of how to use the tables in (VI.18). If one is only interested in the orders of the factor groups $\gamma_i(P)/\gamma_{i+1}(P)$ one can use section VI c).

(VI.17) Table.

$SU_n(K,F)$, $D = K$ with K/k unramified, $n \geq 3$ odd, ($^2A_{n-1}$, quasi-split),

$$F := ((^0 1\,_0))_n, \alpha = n, B = ((^2 1\,_-))_n^\bullet$$

$\overline{n^{n-1}, n-1}$

$SU_n(K,F)$, $D = K$ with K/k ramified, $n \geq 3$ odd, $(n,p) \neq (3,3)$, ($^2A_{n-1}$, quasi-split),

$$F := ((^0 1\,_0))_n, \alpha = n,$$

$$B_1 = \begin{pmatrix} 1 & 1 & (^1 0\,_-) \\ 1 & 1 & - \\ (^1 1\,_-) & - & - \end{pmatrix}^\bullet_{\frac{n-1}{2},1,\frac{n-1}{2}} \qquad B_2 = \begin{pmatrix} 1 & 1 & (^1 1\,_-) \\ 1 & 0 & - \\ (^1 0\,_-) & - & - \end{pmatrix}_{\frac{n-1}{2},1,\frac{n-1}{2}}$$

$\overline{(m+1,m)^m, m, (m,m+1)^m, m}$ for $m = \frac{n-1}{2}$

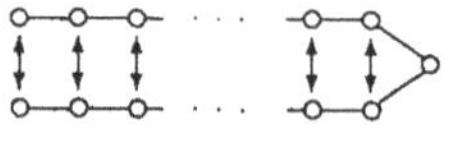

$SU_n(K,F)$, $D = K$ with K/k unramified, $n \geq 4$ even, ($^2A_{n-1}$, quasi-split),

$$F := ((^0 1\,_0))_n, \alpha = n, B = ((^2 1\,_-))_n^\bullet$$

$\overline{n^{n-1}, n-1}$

$SU_n(K,F)$, $D = K$ with K/k ramified, $n \geq 4$ even, ($^2A_{n-1}$, quasi-split),

$$F := ((^0 1\,_0))_n, \alpha = n-1, n_{\frac{n}{2}-1} = 2,$$

$$B_1 = \begin{pmatrix} 1 & 2 & (^1 0\,_-) \\ 2 & 3 & - \\ (^1 1\,_-) & - & - \end{pmatrix}^\bullet_{\frac{n-2}{2},1,\frac{n-2}{2}} \qquad B_2 = \begin{pmatrix} 1 & 2 & (^1 1\,_-) \\ 2 & 1 & - \\ (^1 0\,_-) & - & - \end{pmatrix}_{\frac{n-2}{2},1,\frac{n-2}{2}}$$

$\overline{(\frac{n}{2}+1, \frac{n}{2})^{n-1}}$

$SU_n(K,F)$, $D = K$ with K/k unramified, $n \geq 4$ even, $({}^2A_{n-1})$,

$$F := \begin{pmatrix} ({}^{0}1_{0}) & 0 \\ 0 & \pi \end{pmatrix}_{n-1,1}, \alpha = n, B = \begin{pmatrix} ({}^{2}1_{-}) & 2 \\ - & 1 \end{pmatrix}^{\bullet}_{n-1,1}$$

$\overline{n^{n-1}, n-1}$

$SU_n(K,F)$, $D = K$ with K/k ramified, $n \geq 4$ even, ε unit in k, $-\varepsilon \notin \mathrm{nr}_{K/k}(K)$, $({}^2A_{n-1})$,

$$F := \begin{pmatrix} 0 & 0 & 0 & ({}^{0}1_{0}) \\ 0 & \varepsilon & 0 & 0 \\ 0 & 0 & 1 & 0 \\ ({}^{0}1_{0}) & 0 & 0 & 0 \end{pmatrix}_{\frac{n-2}{2},1,1,\frac{n-2}{2}}, \alpha = n-1, n_{\frac{n-2}{2}} = 2,$$

$$B_1 = \begin{pmatrix} 1 & 2 & ({}^{1}0_{-}) \\ 2 & 3 & - \\ ({}^{1}1_{-}) & - & - \end{pmatrix}^{\bullet}_{\frac{n-2}{2},1,\frac{n-2}{2}}, B_2 = \begin{pmatrix} 1 & 2 & ({}^{1}1_{-}) \\ 2 & 1 & - \\ ({}^{1}0_{-}) & - & - \end{pmatrix}_{\frac{n-2}{2},1,\frac{n-2}{2}}$$

$\overline{(\frac{n}{2}+1, \frac{n}{2})^{n-1}}$

$SO_n(k,F)$, $D = k$ with n odd, $(B_{\frac{n-1}{2}}$, split$)$,

$$F := \begin{pmatrix} ({}^{0}1_{0}) & 0 \\ 0 & \pi \end{pmatrix}_{n-1,1}, \alpha = n-1, n_{\frac{n-3}{2}} = 2,$$

$$B = \begin{pmatrix} 1 & 2 & ({}^{1}0_{-}) & 1 \\ 2 & 1 & - & 2 \\ ({}^{1}0_{-}) & - & - & 1 \\ - & - & - & 0 \end{pmatrix}_{\frac{n-3}{2},1,\frac{n-3}{2},1}$$

$\overline{(\frac{n+1}{2}, \frac{n-1}{2})^{\frac{n-1}{2}}}$

$SO_n(k,F)$, $D=k$ with n odd, ε a unit in k, $-\varepsilon \notin k^{*2}$, $(B_{\frac{n-1}{2}})$

$$F := \begin{pmatrix} 0 & 0 & 0 & ({}^{0}1_{0}) & 0 \\ 0 & 1 & 0 & 0 & 0 \\ 0 & 0 & \varepsilon & 0 & 0 \\ ({}^{0}1_{0}) & 0 & 0 & 0 & 0 \\ 0 & 0 & 0 & 0 & \pi \end{pmatrix}_{\frac{n-3}{2},1,1,\frac{n-3}{2},1} \quad , \alpha = n-1, n_{\frac{n-3}{2}} = 2,$$

$$B = \begin{pmatrix} 1 & 2 & ({}^{1}0\,\_) & 1 \\ 2 & 1 & - & 2 \\ ({}^{1}0\,\_) & - & - & 1 \\ - & - & - & 0 \end{pmatrix}_{\frac{n-3}{2},1,\frac{n-3}{2},1}$$

$\overline{(\frac{n+1}{2},\frac{n-1}{2})^{\frac{n-1}{2}}}$

$Sp_n(k,F)$, $D=k$ with n even, $(C_{\frac{n}{2}}$, split),

$$F := \begin{pmatrix} 0 & ({}^{0}-1_{0}) \\ ({}^{0}1_{0}) & 0 \end{pmatrix}_{\frac{n}{2},\frac{n}{2}} \quad , \alpha = n, B = (({}^{1}1\,\_))_n$$

$\overline{(m+1,m)^m}$ for $m=\frac{n}{2}$

$SU_n(Q,F)$, $D=Q$ with n even, (C_n),

$$F := ((^{0}1_{0}))_n, \alpha = n,$$

$$B_1 = \begin{pmatrix} 2 & (^{2}1_{-}) \\ (^{2}2_{-}) & - \end{pmatrix}_{\frac{n}{2},\frac{n}{2}}, B_2 = \begin{pmatrix} 2 & (^{2}2_{-}) \\ (^{2}1_{-}) & - \end{pmatrix}_{\frac{n}{2},\frac{n}{2}}$$

$\overline{(n+1,n)^n}$

$SU_n(Q,F)$, $D=Q$ with n odd, (C_n),

$$F := ((^{0}1_{0}))_n, \alpha = n,$$

$$B_1 = \begin{pmatrix} 2 & 2 & (^{2}1_{-}) \\ 2 & 2 & - \\ (^{2}2_{-}) & - & - \end{pmatrix}_{\frac{n-1}{2},1,\frac{n-1}{2}}, B_2 = \begin{pmatrix} 2 & 2 & (^{2}2_{-}) \\ 2 & 1 & - \\ (^{2}1_{-}) & - & - \end{pmatrix}_{\frac{n-1}{2},1,\frac{n-1}{2}}$$

$\overline{(n+1,n)^n}$

$SO_n(k,F)$, $D=k$ with n even, (${}^1D_{\frac{n}{2}}$, split),

$$F := \begin{pmatrix} \left({}^{0}1{}_{0}\right) & 0 \\ 0 & \left({}^{0}\pi{}_{0}\right) \end{pmatrix}_{n-2,2}, \alpha = n-2, n_{\frac{n-4}{2}} = 2, n_{n-3} = 2,$$

$$B = \begin{pmatrix} 1 & 2 & \left({}^{1}0{}_{-}\right) & 2 \\ 2 & 1 & - & 4 \\ \left({}^{1}0{}_{-}\right) & - & - & 2 \\ - & - & - & 1 \end{pmatrix}_{\frac{n-4}{2},1,\frac{n-4}{2},1}$$

$(m+1,m)^{\frac{m-2}{2}}, m+2, (m,m+1)^{\frac{m-2}{2}}, m$ for $m := \frac{n}{2}$ and m even

$(m+1,m)^{\frac{m-3}{2}}, (m+1)^2, (m+1,m)^{\frac{m-1}{2}}$ for $m := \frac{n}{2}$ and m odd

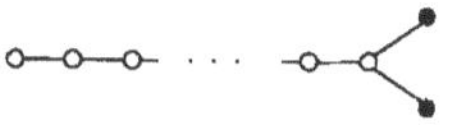

$SO_n(k,F)$, $D=k$ with n even, ε a unit in k, $-\varepsilon \notin k^{*2}$, (${}^1D_{\frac{n}{2}}$),

$$F := \begin{pmatrix} 0 & 0 & \left({}^{0}1{}_{0}\right) & 0 \\ 0 & \left({}^{1}0{}_{\varepsilon}\right) & 0 & 0 \\ \left({}^{0}1{}_{0}\right) & 0 & 0 & 0 \\ 0 & 0 & 0 & \left({}^{\pi}0{}_{\pi\varepsilon}\right) \end{pmatrix}_{\frac{n-4}{2},2,\frac{n-4}{2},2}, \alpha = n-2, n_{\frac{n-4}{2}} = 2, n_{n-3} = 2$$

$$B = \begin{pmatrix} 1 & 2 & \left({}^{1}0{}_{-}\right) & 2 \\ 2 & 1 & - & 4 \\ \left({}^{1}0{}_{-}\right) & - & - & 2 \\ - & - & - & 1 \end{pmatrix}_{\frac{n-4}{2},1,\frac{n-4}{2},1}$$

$(m+1,m)^{\frac{m-2}{2}}, m+2, (m,m+1)^{\frac{m-2}{2}}, m$ for $m := \frac{n}{2}$ and m even

$(m+1,m)^{\frac{m-3}{2}}, (m+1)^2, (m+1,m)^{\frac{m-1}{2}}$ for $m := \frac{n}{2}$ and m odd

$SU_n(Q,F)$, $D=Q$ with n even, F skew Hermitian, $(\,{}^1D_n\,)$,

$$F := \begin{pmatrix} 0 & \left({}^{0}\,{-1}\,{}_{0}\right) & 0 \\ \left({}^{0}\,1\,{}_{0}\right) & 0 & 0 \\ 0 & 0 & \left({}^{\pi_Q}\,0\,{}_{-\pi_Q}\right) \end{pmatrix}_{\frac{n-2}{2},\frac{n-2}{2},2}, \alpha = n-1, n_{n-2}=2,$$

$$B_1 = \begin{pmatrix} 2 & \left({}^{2}\,1\,{}_{-}\right) & 4 \\ \left({}^{2}\,0\,{}_{-}\right) & - & 4 \\ - & - & 4 \end{pmatrix}_{\frac{n-2}{2},\frac{n-2}{2},1}, B_2 = \begin{pmatrix} 2 & \left({}^{2}\,0\,{}_{-}\right) & 4 \\ \left({}^{2}\,1\,{}_{-}\right) & - & 4 \\ - & - & 2 \end{pmatrix}_{\frac{n-2}{2},\frac{n-2}{2},1}$$

$(n+1,n)^{\frac{n-2}{2}}, n+2, (n,n+1)^{\frac{n-2}{2}}, n$

$SU_n(Q,F)$, $D=Q$ with n odd, F skew Hermitian, $(\,{}^1D_n\,)$,

$$F := \begin{pmatrix} 0 & 0 & \left({}^{0}\,{-1}\,{}_{0}\right) & 0 \\ 0 & q & 0 & 0 \\ \left({}^{0}\,1\,{}_{0}\right) & 0 & 0 & 0 \\ 0 & 0 & 0 & \left({}^{q\pi_Q}\,0\,{}_{\pi_Q}\right) \end{pmatrix}_{\frac{n-3}{2},1,\frac{n-3}{2},2}, \alpha = n-1, n_{n-2}=2,$$

$$B_1 = \begin{pmatrix} 2 & 2 & \left({}^{2}\,1\,{}_{-}\right) & 4 \\ 2 & 0 & - & 4 \\ \left({}^{2}\,0\,{}_{-}\right) & - & - & 4 \\ - & - & - & 4 \end{pmatrix}_{\frac{n-3}{2},1,\frac{n-3}{2},1}, B_2 = \begin{pmatrix} 2 & 2 & \left({}^{2}\,0\,{}_{-}\right) & 4 \\ 2 & 1 & - & 4 \\ \left({}^{2}\,1\,{}_{-}\right) & - & - & 4 \\ - & - & - & 2 \end{pmatrix}_{\frac{n-3}{2},1,\frac{n-3}{2},1}$$

$(n+1,n)^{\frac{n-3}{2}}, (n+1)^2, (n+1,n)^{\frac{n-1}{2}}$

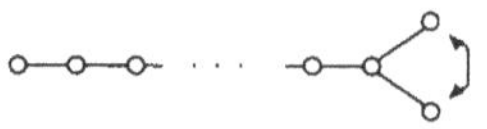

$SO_n(k,F)$, $D = k$ with n even, (${}^2D_{\frac{n}{2}}$, quasi-split, two possibilities: ramified splitting field),

$$F := \begin{pmatrix} \left({}^{0}1_{0}\right) & 0 \\ 0 & \pi \end{pmatrix}_{n-1,1}, \alpha = n, B = \begin{pmatrix} \left({}^{1}0_{-}\right) & 1 \\ - & 0 \end{pmatrix}_{n-1,1}$$

$\overline{(\frac{n}{2}, \frac{n}{2} - 1)^{\frac{n}{2}}}$

$SO_n(k,F)$, $D = k$ with n even, ε a unit in k, $-\varepsilon \notin k^{*2}$, (${}^2D_{\frac{n}{2}}$ unramified splitting field),

$$F := \begin{pmatrix} 0 & 0 & \left({}^{0}1_{0}\right) & 0 \\ 0 & \left({}^{1}0_{\varepsilon}\right) & 0 & 0 \\ \left({}^{0}1_{0}\right) & 0 & 0 & 0 \\ 0 & 0 & 0 & \left({}^{0}\pi_{0}\right) \end{pmatrix}_{\frac{n-4}{2},2,\frac{n-4}{2},2}, \alpha = n-2, n_{\frac{n-4}{2}} = 2, n_{n-3} = 2$$

$$B = \begin{pmatrix} 1 & 2 & \left({}^{1}0_{-}\right) & 2 \\ 2 & 1 & - & 4 \\ \left({}^{1}0_{-}\right) & - & - & 2 \\ - & - & - & 1 \end{pmatrix}_{\frac{n-4}{2},1,\frac{n-4}{2},1}$$

$\overline{(m+1,m)^{\frac{m-2}{2}}, m+2, (m,m+1)^{\frac{m-2}{2}}, m}$ for $m := \frac{n}{2}$ and m even

$\overline{(m+1,m)^{\frac{m-3}{2}}, (m+1)^2, (m+1,m)^{\frac{m-1}{2}}}$ for $m := \frac{n}{2}$ and m odd

$SU_n(Q,F)$, $D=Q$ with n even, F skew Hermitian, (${}^2D_n$),

$$F_1 := \begin{pmatrix} 0 & ({}^{0}{-1}_{0}) & 0 \\ ({}^{0}1_{0}) & 0 & 0 \\ 0 & 0 & ({}^{q\pi_Q}0_{\pi_Q}) \end{pmatrix}_{\frac{n-2}{2},\frac{n-2}{2},2}, \alpha = n-1, n_{n-2} = 2,$$

$$B_1 = \begin{pmatrix} 2 & ({}^{2}1_{-}) & 4 \\ ({}^{2}0_{-}) & - & 4 \\ - & - & 4 \end{pmatrix}_{\frac{n-2}{2},\frac{n-2}{2},1}, B_2 = \begin{pmatrix} 2 & ({}^{2}0_{-}) & 4 \\ ({}^{2}1_{-}) & - & 4 \\ - & - & 2 \end{pmatrix}_{\frac{n-2}{2},\frac{n-2}{2},1}$$

$\overline{(n+1,n)^{\frac{n-2}{2}}, n+2, (n,n+1)^{\frac{n-2}{2}}, n}$

$SU_n(Q,F)$, $D=Q$ with n even, F skew Hermitian, (${}^2D_n$, two groups),

$$F_2 := \begin{pmatrix} 0 & 0 & ({}^{0}{-1}_{0}) & 0 \\ 0 & q & 0 & 0 \\ ({}^{0}1_{0}) & 0 & 0 & 0 \\ 0 & 0 & 0 & \pi_Q \end{pmatrix}_{\frac{n-2}{2},1,\frac{n-2}{2},1} \quad \alpha = n \quad \text{or}$$

$$F_3 := \begin{pmatrix} 0 & 0 & ({}^{0}{-1}_{0}) & 0 \\ 0 & q & 0 & 0 \\ ({}^{0}1_{0}) & 0 & 0 & 0 \\ 0 & 0 & 0 & q\pi_Q \end{pmatrix}_{\frac{n-2}{2},1,\frac{n-2}{2},1} \quad \alpha = n,$$

$$B_1 = \begin{pmatrix} 2 & 2 & ({}^{2}1_{-}) & 2 \\ 2 & 0 & - & 2 \\ ({}^{2}0_{-}) & - & - & 2 \\ - & - & - & 1 \end{pmatrix}_{\frac{n-2}{2},1,\frac{n-2}{2},1}, B_2 = \begin{pmatrix} 2 & 2 & ({}^{2}0_{-}) & 2 \\ 2 & 1 & - & 2 \\ ({}^{2}1_{-}) & - & - & 2 \\ - & - & - & 0 \end{pmatrix}_{\frac{n-2}{2},1,\frac{n-2}{2},1}$$

$\overline{(n,n-1)^n}$

$SU_n(Q,F)$, $D = Q$ with n odd, F_1 skew Hermitian, $({}^2D_n)$,

$$F := \begin{pmatrix} 0 & 0 & ({}^{0}\,{-1}\,{}_{0}) & 0 & 0 \\ 0 & q & 0 & 0 & 0 \\ ({}^{0}\,1\,{}_{0}) & 0 & 0 & 0 & 0 \\ 0 & 0 & 0 & 0 & \pi_Q \\ 0 & 0 & 0 & -\pi_Q & 0 \end{pmatrix}_{\frac{n-3}{2},1,\frac{n-3}{2},1,1}, \alpha = n-1, n_{n-2} = 2,$$

$$B_1 = \begin{pmatrix} 2 & 2 & ({}^{2}\,1\,{}_{-}) & 4 \\ 2 & 0 & - & 4 \\ ({}^{2}\,0\,{}_{-}) & - & - & 4 \\ - & - & - & 4 \end{pmatrix}_{\frac{n-3}{2},1,\frac{n-3}{2},1}, B_2 = \begin{pmatrix} 2 & 2 & ({}^{2}\,0\,{}_{-}) & 4 \\ 2 & 1 & - & 4 \\ ({}^{2}\,1\,{}_{-}) & - & - & 4 \\ - & - & - & 2 \end{pmatrix}_{\frac{n-3}{2},1,\frac{n-3}{2},1}$$

$\overline{(n+1,n)^{\frac{n-3}{2}}, (n+1)^2, (n+1,n)^{\frac{n-1}{2}}}$

$SU_n(Q,F)$, $D = Q$ with n odd, F skew Hermitian, $({}^2D_n$, two groups),

$$F_2 := \begin{pmatrix} 0 & ({}^{0}\,{-1}\,{}_{0}) & 0 \\ ({}^{0}\,1\,{}_{0}) & 0 & 0 \\ 0 & 0 & \pi_Q \end{pmatrix}_{\frac{n-2}{2},\frac{n-2}{2},1}, \text{ or } F_3 := \begin{pmatrix} 0 & ({}^{0}\,{-1}\,{}_{0}) & 0 \\ ({}^{0}\,1\,{}_{0}) & 0 & 0 \\ 0 & 0 & q\pi_Q \end{pmatrix}_{\frac{n-2}{2},\frac{n-2}{2},1}$$

$\alpha = n$,

$$B_1 = \begin{pmatrix} 2 & ({}^{2}\,1\,{}_{-}) & 2 \\ ({}^{2}\,0\,{}_{-}) & - & 2 \\ - & - & 1 \end{pmatrix}_{\frac{n-1}{2},\frac{n-1}{2},1}, B_2 = \begin{pmatrix} 2 & ({}^{2}\,0\,{}_{-}) & 2 \\ ({}^{2}\,1\,{}_{-}) & - & 2 \\ - & - & 0 \end{pmatrix}_{\frac{n-1}{2},\frac{n-1}{2},1}$$

$\overline{(n,n-1)^n}$

Here is an example of how to use the tables

(VI.18) Example.
(i) One extracts information about topological generators and about the lower central series of a Sylow pro-p-subgroup P of $SU_5(K,F)$ with K/k ramified of degree 2 (type ${}^2A_4$, quasi-split) from the table in the following way.
The Gram-matrix F of the ε-Hermitian form is given as

$$F = \begin{pmatrix} 0 & 0 & 0 & 0 & 1 \\ 0 & 0 & 0 & 1 & 0 \\ 0 & 0 & 1 & 0 & 0 \\ 0 & 1 & 0 & 0 & 0 \\ 1 & 0 & 0 & 0 & 0 \end{pmatrix}.$$

An involution $^\circ : K^{5\times 5} \to K^{5\times 5} : x \to F\overline{x}^{tr}F^{-1}$ is defined where $\bar{}$ denotes the Galois automorphism of K over k. Then $SU_5(K,F) = \{x \in K^{5\times 5} | xx^\circ = 1, \det(x) = 1\}$. The ring of integers of K is denoted by $\mathcal{O}$ and the uniformising element is Π. The ring of integers of k is denoted by o and the uniformising element is π. A minimal $^\circ$-invariant hereditary order Λ has exponent matrix $\begin{pmatrix} 0 & 0 & 0 & 0 & 0 \\ 1 & 0 & 0 & 0 & 0 \\ 1 & 1 & 0 & 0 & 0 \\ 1 & 1 & 1 & 0 & 0 \\ 1 & 1 & 1 & 1 & 0 \end{pmatrix}$. This means that every element $(a_{ij})_{1\le i,j\le 5} \in \Lambda$ is of the form $\begin{cases} a_{ij} \in \mathcal{O} & \text{for } i \le j \\ a_{ij} \in \Pi\mathcal{O} & \text{for } i > j \end{cases}$. The radical rad$\Lambda$ has exponent matrix $\begin{pmatrix} 1 & 0 & 0 & 0 & 0 \\ 1 & 1 & 0 & 0 & 0 \\ 1 & 1 & 1 & 0 & 0 \\ 1 & 1 & 1 & 1 & 0 \\ 1 & 1 & 1 & 1 & 1 \end{pmatrix}$.
The set $\{x \in \text{rad}\Lambda | x^\circ = -x\}$ is denoted by $(\text{rad}\Lambda)^-$. One observes from the table that $\alpha = 5$. This means that $\Lambda/\text{rad}\Lambda$ is the direct product of α simple $o/\pi o$-algebras; here $\Lambda/\text{rad}\Lambda = (\mathcal{O}/\Pi\mathcal{O})^5$. It follows from (VI.13) that $P = (\text{rad}\Lambda)^- C_{LP} \cap SU$ where C_{LP} is induced by the Cayley map. Therefore, to obtain generators for the group, one needs (Lie-) generators for $(\text{rad}\Lambda)^-$ which are of the form $d_1(a_1,\dots,a_n)$ with $a_1,\dots,a_n \in K$ satisfying $F(\overline{d_1(a_1,\dots,a_n)})^{tr}F^{-1} = -d_1(a_1,\dots,a_n)$. This condition imposes linear equations on the entries $a_1,\dots,a_n$. One applies the Cayley map c_{LP} : $x \mapsto (1-x)(1+x)^{-1}$ to these elements. Specifying the field k to be $\mathbb{Q}_p$, one can generate $SA \cap (\text{rad}\Lambda)^-$ by

$$d_1(1,0,0,-1,0) = \begin{pmatrix} 0 & 1 & 0 & 0 & 0 \\ 0 & 0 & 0 & 0 & 0 \\ 0 & 0 & 0 & 0 & 0 \\ 0 & 0 & 0 & 0 & -1 \\ 0 & 0 & 0 & 0 & 0 \end{pmatrix},\ d_1(0,1,-1,0,0),\ d_1(0,0,0,0,\Pi)$$

as Lie lattice over $\mathbb{Z}_p$, where $\Pi^2 = p$ or $\Pi^2 = \varepsilon p$ for some non-square unit $\varepsilon \in \mathbb{Z}_p^*$, depending on the choice of K. Topological generators for the group are the images of these generators of $(\mathrm{rad}\Lambda)^-$ under the Cayley map c_{LP}. These are

$$\begin{pmatrix} 1 & -2 & 0 & 0 & 0 \\ 0 & 1 & 0 & 0 & 0 \\ 0 & 0 & 1 & 0 & 0 \\ 0 & 0 & 0 & 1 & 2 \\ 0 & 0 & 0 & 0 & 1 \end{pmatrix}, \begin{pmatrix} 1 & 0 & 0 & 0 & 0 \\ 0 & 1 & -2 & -2 & 0 \\ 0 & 0 & 1 & 2 & 0 \\ 0 & 0 & 0 & 1 & 0 \\ 0 & 0 & 0 & 0 & 1 \end{pmatrix}, \begin{pmatrix} 1 & 0 & 0 & 0 & 0 \\ 0 & 1 & 0 & 0 & 0 \\ 0 & 0 & 1 & 0 & 0 \\ 0 & 0 & 0 & 1 & 0 \\ -2\Pi & 0 & 0 & 0 & 1 \end{pmatrix}.$$

The dimension matrices B_r can be used to read off the orders of the factors of the lower central series of P since $\mathrm{rad}(\Lambda)^-$ is saturated, cf. (VI.9).
These matrices, given in the table, read as

$$B_1 = (b_{s,t}^{(1)})_{0 \le s,t \le \alpha-1} = \begin{pmatrix} 1 & 1 & 1 & 1 & 0 \\ 1 & 1 & 1 & 0 & - \\ 1 & 1 & 1 & - & - \\ 1 & 1 & - & - & - \\ 1 & - & - & - & - \end{pmatrix}^{\bullet}, \quad B_2 = (b_{s,t}^{(1)}) = \begin{pmatrix} 1 & 1 & 1 & 1 & 1 \\ 1 & 1 & 1 & 1 & - \\ 1 & 1 & 0 & - & - \\ 1 & 0 & - & - & - \\ 0 & - & - & - & - \end{pmatrix}$$

The dimension matrices $B_r = (b_{s,t}^{(r)})_{0 \le s,t \le \alpha}$ for $r \ge 2$ are defined as $B_{1+2k} = B_1$ and $B_{2k} = B_2$ for $k \in \mathbb{N}$. The entries $b_{ij}^{(r)} = -$ count as 0. As defined in (VI.16) one has $b_i = \sum_{t-s \equiv i} b_{s,t}^{(r)}$ where $s, t \in \{0, 1, \dots, \alpha - 1\}$ and $\alpha(r-1) \le i < \alpha r$ and all indices taken modulo α. The sequence $(b_1, b_2, \dots)$ is thus given by $3, 2, 3, 2, 3, 2, 3, 2, 3, 2, \dots$, in particular $3 = b_5 = b_{15} = b_{25} = \dots$. But this value must be reduced by 1 as it is indicated by $\bullet$ above the matrix B_1. The reason for this is that, if $i = 5 \bmod 10$, then $\gamma_i(P)/\gamma_{i+1}(P)$ is generated by diagonal matrices, and the determinant condition reduces the dimension by 1. The orders of the sections of the lower central series of P for a general p-adic field k with inertia degree f are $p^{3f}, p^{2f}, p^{3f}, p^{2f}, p^{2f}, p^{2f}, p^{3f}, p^{2f}, p^{3f}, p^{2f}$ after which the pattern repeats.

(ii) Consider the Sylow pro-3-subgroup of $SO_5(\mathbb{Q}_3, F)$ (type B_2, split) with

$$F = \left(\begin{array}{c|cc|c|c} 0 & 0 & 0 & 1 & 0 \\ \hline 0 & 0 & 1 & 0 & 0 \\ 0 & 1 & 0 & 0 & 0 \\ \hline 1 & 0 & 0 & 0 & 0 \\ \hline 0 & 0 & 0 & 0 & 3 \end{array}\right).$$

The involution defining $SO_5(\mathbb{Q}_3, F)$ is given by $^\circ : \mathbb{Q}_3^{5\times5} \to \mathbb{Q}_3^{5\times5} : x \mapsto F x^{tr} F^{-1}$. Let Λ be an associated $^\circ$-invariant minimal hereditary order Λ. The number of simple components of the $\Lambda/\mathrm{rad}\Lambda$ is $\alpha = 4$ and $n_1 = 2$. Therefore $(\mathrm{rad}\Lambda)^-$ is generated by

$$d_1((1,0),(0,-1)^{tr},0,0) = \left(\begin{array}{c|cc|c|c} 0 & 1 & 0 & 0 & 0 \\ \hline 0 & 0 & 0 & 0 & 0 \\ 0 & 0 & 0 & -1 & 0 \\ \hline 0 & 0 & 0 & 0 & 0 \\ \hline 0 & 0 & 0 & 0 & 0 \end{array}\right),$$

$$d_1((0,1),(-1,0)^{tr},0,0), d_1((0,0),(0,0)^{tr},1,-3).$$

Therefore topological generators for the Sylow pro-p-subgroup of $SO_5(\mathbb{Q}_3, F)$ are

$$\left(\begin{array}{c|cc|c|c} 1 & -2 & 0 & 0 & 0 \\ \hline 0 & 1 & 0 & 0 & 0 \\ 0 & 0 & 1 & 2 & 0 \\ \hline 0 & 0 & 0 & 1 & 0 \\ \hline 0 & 0 & 0 & 0 & 1 \end{array}\right), \left(\begin{array}{c|cc|c|c} 1 & 0 & -2 & 0 & 0 \\ \hline 0 & 1 & 0 & 2 & 0 \\ 0 & 0 & 1 & 0 & 0 \\ \hline 0 & 0 & 0 & 1 & 0 \\ \hline 0 & 0 & 0 & 0 & 1 \end{array}\right), \left(\begin{array}{c|cc|c|c} 1 & 0 & 0 & 0 & 0 \\ \hline 0 & 1 & 0 & 0 & 0 \\ 0 & 0 & 1 & 0 & 0 \\ \hline -6 & 0 & 0 & 1 & -2 \\ \hline 6 & 0 & 0 & 0 & 1 \end{array}\right).$$

The lower central series has the pattern $3^3, 3^2, 3^3, 3^2$ after which the pattern repeats.

Proof of the claims in the tables.

(i) Completeness of enumeration of groups.
Cf. [Sat 71] for the classification of simple algebraic groups over $\mathbb{Q}_p$ up to isogeny. For each Dynkin diagram one is left with the problem of enumerating the Hermitian, bilinear or skew-Hermitian forms in each case, cf. [Tsu 61] for the skew-Hermitian case over quaternion algebras and [Scha 85] for the other cases.

(ii) Maximality of the $SU(\Lambda)$ as pro-p-subgroups of SU.
By (VI.8) a maximal pro-p-subgroup of SU is of the form $(1 + \mathrm{rad}\Lambda) \cap SU = (\mathrm{rad}\Lambda)^{-}C_{LP}$ for some hereditary °-invariant order Λ in A. Since $\mathrm{rad}\Lambda \subset \mathrm{rad}\Lambda_1$ if and only if $\Lambda_1 \subset \Lambda$ for hereditary orders, one is done in the case Λ is a minimal hereditary order. In the other cases one quickly convinces oneself that the °-invariant hereditary orders Γ contained in Λ yield $(\mathrm{rad}\Gamma)^{-} = (\mathrm{rad}\Lambda)^{-}$. It is easy to see that our choices of Λ make $(\mathrm{rad}\Lambda)^{-}$ semi-saturated in all cases, even for $SU_3(K)$, K/k ramified, $p = 3$.

(iii) Lower central series of $SU(\Lambda)$.
We only have to prove that $(\mathrm{rad}\Lambda)^{-}$ is saturated in all cases except $SU_3(K)$, K/k ramified, $p = 3$.

Prior to the discussion of the general case, let us remark that with the help of Lemma (VI.11) or with a symbolic manipulation package like MAPLE, one can check any of the cases of (VI.17) for a given degree n with K arbitrary. Let us do $SU_3(K)$ with K/k ramified as an example. We arrange the elements of $(\mathrm{rad}\Lambda)^{-}$ according to the layers $((\mathrm{rad}\Lambda)^{i})^{-}/((\mathrm{rad}\Lambda)^{i+1})^{-}$:

$$\begin{aligned}
&a_1 = d_1(1, -1, 0),\ a_2 = d_1(0, 0, \Pi);\\
&b_1 = [a_1, a_2] = d_2(0, -\Pi, -\Pi);\\
&c_1 = [a_1, b_1] = d_0(-\Pi, 2\Pi, -\Pi),\ ([a_2, b_1] = 0);\\
&d_1 = [a_1, c_1] = d_1(3\Pi, 3\Pi, 0),\ ([a_2, c_1] = 0);\\
&e_1 = [a_1, d_1] = d_2(6\Pi, 0, 0),\ e_2 = [a_2, d_1] = d_2(0, -3\Pi^2, 3\Pi^2);\\
&f_1 = [a_1, e_2] = d_0(-3\Pi^2, 0, 3\Pi^2),\ ([a_1, e_1] = 0,\ [a_2, e_1] = 2f_1,\ [a_2, e_2] = 0);\\
&\ [a_1, f_1] = 3\Pi^2 a_1,\ [a_2, f_1] = -6\Pi^2 a_2.
\end{aligned}$$

Note in the third layer there is just one element, since the trace must be zero, i.e. B_1 has a •. This calculation proves that $(\mathrm{rad}\Lambda)^{-}$ is saturated if and only if $p > 3$.

Passing to the general case, recall from (VI.14) that the involution ° induces a permutation ι on the components $X_0, \ldots, X_{\alpha-1}$ of $\Lambda/\mathrm{rad}\Lambda$ with 0 or 1 or 2 fixed points.

Accordingly the number of fixed points on the set of the components $Y_{0,\iota}^{(i)},\ldots,Y_{\alpha-1,\alpha-1+i}^{(i)}$ follows the pattern $0,2,0,2,\ldots$ or $1,1,1,\ldots$ or $2,0,2,0,\ldots$. In the setup of table (VI.17) the fixed points can be visualised in the same positions where the Gram matrix F has non zero entries if partioned according to Λ. For instance in the first case the permutation ι is given by $(0,\alpha-1)(1,\alpha-2)\cdots(\frac{\alpha}{2}-1,\frac{\alpha}{2}+1)$.
Claim: The Lie lattices of table (VI.17) are saturated, if the first two or three in each series are saturated.
Proof. To prove that $[(\mathrm{rad}\Lambda)^-,((\mathrm{rad}\Lambda)^{i-1})^-]=((\mathrm{rad}\Lambda)^i)^-$, it suffices to see that any component of $((\mathrm{rad}\Lambda)^i)^-/((\mathrm{rad}\Lambda)^{i+1})^-$ can be obtained from a component of $((\mathrm{rad}\Lambda)^{i-1})^-/((\mathrm{rad}\Lambda)^i)^-$ and one of $(\mathrm{rad}\Lambda)^-/((\mathrm{rad}\Lambda)^2)^-$ by taking Lie commutators (of representatives). For involutions of the second kind, where the first dimension matrix B_1 gets a $\bullet$, one has to allow for the fact that traces are zero, i. e. which might force one to take two components rather than one in case $i=\alpha$ or $i-1=\alpha$. The components of $((\mathrm{rad}\Lambda)^i)^-/((\mathrm{rad}\Lambda)^{i+1})^-$ are of the form $C_{s,t}^{(i)}:=(Y_{s,t}^{(i)}\oplus Y_{t\iota,s\iota}^{(i)})^-(=C_{t\iota,s\iota}^{(i)})$ with $t-s\equiv i \mod \alpha$ and $s\neq t\iota$ or $C_{s,s\iota}^{(i)}:=(Y_{s,s\iota}^{(i)})^-$ with $s\iota-s\equiv i \mod\alpha$. In the first case one would have to look at commutators of $C_{s,t-1}^{(i-1)}$ with $C_{t-1,t}^{(1)}$ or of $C_{s-1,t}^{(i-1)}$ with $C_{s-1,s}^{(1)}$. Note, one of the two $C_{s,t-1}^{(i-1)}$ or $C_{s-1,t}^{(i-1)}$ might be 0, in which case one is forced to take the other one. But note also that none of the $C_{s,s+1}^{(i)}$ is zero, which is very important and forces the choice of Λ as it stands. The other problem is that Lie commutators do not just induce $o/\pi o$-bilinear maps

$$C_{s,t-1}^{(i-1)}\times C_{t-1,t}^{(1)}\to C_{s,t}^{(i)}\quad\text{or}\quad C_{s-1,s}^{(1)}\times C_{s-1,t}^{(i-1)}\to C_{s,t}^{(i)}$$

whose surjectivity one has to check, but that the image involves two rather than one component by chance. (This actually must happen on the diagonal because of the trace 0 condition for involutions of the second kind.) The key idea is now that both issues on these bilinear maps can be checked in a smaller environment, i.e. for a Lie lattice of essentially the same type with smaller n, provided n is beyond a certain size. To be more specific, write $\Lambda=\bigoplus_{i,j=0}^{\alpha-1}\Lambda_{ij}$, where $\Lambda_{00},\ldots,\Lambda_{\alpha-1,\alpha-1}$ are maximal orders (covering the components of $\Lambda/\mathrm{rad}\Lambda$) and Λ_{ij} are Λ_{ii}-Λ_{jj}-bimodules such that the involution $^\circ$ of A permutes the Λ_{ij} according to ι, i.e. $\Lambda_{ij}^\circ=\Lambda_{j\iota,i\iota}$. That this is possible in all cases one immediately sees from the simple form of F in (VI.17). To pass from Λ to a hopefully smaller Λ' to check the above statements on $C_{s,t}^{(i)}$ one chooses $\Lambda'=\bigoplus_{ij\in S}\Lambda_{ij}$ where S is chosen so that $(x,y)\in S\Rightarrow(y\iota,x\iota),(x,x),(y,x)\in S$ to make Λ' an hereditary order with involution and $(s,t)\in S$ together with any of the above choices $(s-1,t)$ and $(s-1,s)$ or $(s,t-1)$ and $(t-1,t)$ to have it represent the proper bilinear map.
Applying this process once or twice respectively its version with a slightly more generous choice for S will reduce the n and if $n\geq n_0+1$ where n_0 is the smallest possible degree to yield a simple algebraic group. This proves the claim.
To finish the proof one only has to investigate the first two Lie lattices of each series either by hand and (VI.11) or by computer. q.e.d.

In conclusion of this chapter, remark that in most cases, where the characteristic of K is zero the groups $SU(\Lambda)/Z(SU(\Lambda))$ are maximal $\tilde{p}$-groups with the obvious exception where $K/\mathbb{Q}_p$ has automorphisms of p-power order and the less obvious exceptions in the case of involutions of the second kind, where rational automorphisms might turn up or if the group has a root system of type D_4 and a diagram automorphism might turn up.

c) Table of patterns for lower central series

The pattern of the lower central series is encoded in the following way.
For each factor group γ_i/γ_{i+1} the number of copies of cyclic groups C_p divided by f is given where f is the inertia degree of k. A sequence of numbers enclosed in brackets and raised to some power x means the repetition of this sequence x-times. The whole sequence is overlined to indicate that it repeats itself. E.g. $\overline{n^{n-1}, n-1}$ means $\gamma_1/\gamma_2 \cong C_p^{fn}, \dots, \gamma_{n-1}/\gamma_n \cong C_p^{fn}, \gamma_n/\gamma_{n+1} \cong C_p^{f(n-1)}$ and so on.

(VI.19) Table.

$\overline{n^{n-1}, n-1}$

$^1A_{n-1}$ split, $n \geq 2$, $p \nmid n$, cf. Chapter V [1]
$^2A_{n-1}$ quasi-split, unramified splitting field
$^2A_{n-1}$ non-split, unramified splitting field, n even

$\overline{(m+1, m)^m, m, (m, m+1)^m, m}$ for $m = \frac{n-1}{2}$

$^2A_{n-1}$ quasi-split, ramified splitting field, n odd

$\overline{(\frac{n}{2}+1, \frac{n}{2})^{n-1}}$

$^2A_{n-1}$ quasi-split, ramified splitting field, n even
$^2A_{n-1}$ non-split, ramified splitting field, n even

[1]The same holds probably for the non-split groups, but details have not yet been checked.

$\overline{(n+1,n)^n}$

B_n split
B_n non-split
C_n split
C_n non-split

$\overline{(n+1,n)^{\frac{n-2}{2}}, n+2, (n,n+1)^{\frac{n-2}{2}}, n}$

${}^1D_n$ split, n even
${}^1D_n$ non-split, n even, F orthogonal
${}^1D_n$ non-split, n even, F skew Hermitian form over quaternion algebra
${}^2D_n$ quasi-split, n even, unramified splitting field
${}^2D_n$ non-split, n even, F_1 skew Hermitian form over quaternion algebra

$\overline{(n+1,n)^{\frac{n-3}{2}}, (n+1)^2, (n+1,n)^{\frac{n-1}{2}}}$

${}^1D_n$ split, n odd
${}^1D_n$ non-split, n odd, F orthogonal
${}^1D_n$ non-split, n odd, F skew Hermitian form over quaternion algebra
${}^2D_n$ quasi-split, n odd, unramified splitting field
${}^2D_n$ non-split, n odd, F_1 skew Hermitian form over quaternion algebra

$\overline{(n,n-1)^n}$

${}^2D_n$ quasi-split, ramified splitting field
${}^2D_n$ non-split, F_2, F_3 skew Hermitian form over quaternion algebra

VII Some thin groups

In this section we exhibit a class of insoluble linear pro-p-groups that have average width arbitrarily close to 1.

Let K be either $\mathbb{Q}_p$ or $\mathbb{F}_p((t))$, and let V be a vector space over K of dimension p^α for some $\alpha > 0$. Let $v_1, v_2, \ldots, v_{p^\alpha}$ be a basis for V. For any integer i, define $v_i = \pi^k v_j$ where π is an element of K of value 1, and $i = j + kp^\alpha$, with $1 \leq j \leq p^\alpha$. Let V_i denote the $\mathcal{O}$-module generated by $\{v_j : j \geq i\}$. So V_{i+1} is of index p in V_i for all i. Now let P be the subgroup of $SL(\mathcal{O}, V_1)$ that normalises V_i and centralises V_i/V_{i+1} for all i. Then P is a Sylow pro-p-subgroup of $SL(\mathcal{O}, V_1)$, and is generated by $\{g_1, g_2, \ldots, g_{p^\alpha}\}$, where g_i maps v_i to $v_i + v_{i+1}$ and fixes v_j for $1 \leq j \leq p^\alpha$, $j \neq i$. Let z be the K-automorphism of V that sends v_i to v_{i+1} for all i. Then z^{p^α} is the scalar πI, where I is the identity. Now z normalises P, and we define G to be the subgroup of $PGL(V)$ generated by z and P modulo scalars.

Most of our calculations take place in P. Clearly P has a filtration $P = P_1 > P_2 > \cdots$, where

$$P_i = \cap_{j=1}^{p^\alpha} C_P(V_j/V_{j+1}) = \cap_{j=1}^{\infty} C_P(V_j/V_{j+1}).$$

This filtration is in fact the lower central series of P, though we shall not need this fact.

(VII.1) Lemma. *P_i/P_{i+1} is an elementary abelian group of rank p^α if $i \not\equiv 0 \bmod p^\alpha$, of rank $p^\alpha - 1$ if $i \equiv 0 \bmod p^\alpha$.*

Proof. Let $g \in P_i$, so $v_j g \equiv v_j + a_j v_{j+i} \bmod V_{i+j+1}$. Here the a_j are integers in the range 0 to $p-1$, which we also take to be elements of $\mathbb{F}_p$. Now $g \mapsto (a_1, a_2, \ldots, a_{p^\alpha})$ defines an injection θ_i of P_i/P_{i+1} into $\mathbb{F}_p^{p^\alpha}$. The only restriction on the a_j arises from the fact that we are working in the special linear group, so the determinant of g must be 1. If $i \not\equiv 0 \bmod p^\alpha$ then defining g by $v_j g = v_j + a_j v_{j+i}$ where a_j is non-zero for just one value of j in the range $[1, \ldots, p^\alpha]$ implies that g is a transvection, and hence has determinant 1. So there is no restriction on the a_j if $i \not\equiv 0 \bmod p^\alpha$. If $i \equiv 0 \bmod p^\alpha$ then the condition that g should have determinant 1 implies at once that $a_1 + a_2 + \cdots + a_{p^\alpha} \equiv 0 \bmod p$. This is the only condition on the a_j, as one can define g to have a diagonal matrix, with respect to the basis $\{v_1, v_2, \ldots, v_{p^\alpha}\}$, where one diagonal entry is $b = 1 + a\pi^k$, where $a \neq 0$, and $i = kp^\alpha$, another diagonal entry is b^{-1}, and the other diagonal entries are 1. This completes the proof. q.e.d.

Note that P is a subgroup of $SL(\mathcal{O}, V_1)$, but we are really interested in $PSL(\mathcal{O}, V_1)$. For odd p this will make no difference, as $\mathcal{O}$ has no non-trivial units congruent to 1 modulo (π) of order a power of p in this case. However, if $p = 2$ and we are in characteristic 0, then -1 is such a unit. This implies that if we replace P by its image in the projective linear group, the dimension of $P_{p^\alpha}/P_{p^\alpha+1}$ is one less in this case. A similar complication occurs if we allow K to be a totally ramified extension of $\mathbb{Q}_p$, as 1-units of order a power of p may also arise in this case.

(VII.2) Lemma. *Let A denote the group algebra $\mathbb{F}_p C_{p^\alpha}$ of the cyclic group generated by z, and let $\phi : \mathbb{F}_p^{p^\alpha} \to A$ be defined by $(a_1, a_2, \ldots, a_{p^\alpha}) \mapsto \sum a_j z^{j-1}$. Then $\theta_i\phi$ is a z-module monomorphism, where θ_i is defined as in the proof of Lemma (VII.1).*

Proof. It is easy to see that z acts in both cases by cyclic permutation of the a_j. q.e.d.

(VII.3) Lemma. *The action of G on P_i/P_{i+1} is uniserial. The image of $\theta_i\phi$ is the whole of A if $i \not\equiv 0 \bmod p^\alpha$, and is the augmentation ideal of A otherwise.*

Proof. The first statement follows from the fact that z acts uniserially on A, and the other statements are clear. q.e.d.

If $g \in P_i$, we shall say that g lies in the *i-th layer* of P_i if the image of g modulo P_{i+1} under $\theta_i\phi$ lies in the i-th power of the augmentation ideal of A. For $i = p^\alpha - 1$ we shall say that g lies in the *bottom layer* of P_i, and similarly for the *next to bottom layer*.

We now consider the effect of commutation by the element y of P defined by $v_1 y = v_1 + v_2$, and $v_i y = v_i$ for $1 < i \leq p^\alpha$. Clearly y centralises P_i/P_{i+1} for all i.

(VII.4) Lemma. *Let $g \in P_i$ have image in P_i/P_{i+1} corresponding under $\theta_i\phi$ to $\sum g_j z^{j-1}$. Then $[g, y]$ lies in P_{i+1}, and has image in P_{i+1}/P_{i+2} corresponding under $\theta_{i+1}\phi$ to $g_{p^\alpha+1-i} z^{p^\alpha - i} - g_2$.*

Proof. This is a very simple calculation. q.e.d.

Note that the first layer of P_i is not the top layer unless $i \equiv 0 \bmod p^\alpha$ as there is also a zero-th layer.

(VII.5) Lemma. *Let g lie in the bottom layer of P_i, where $i \not\equiv 0 \bmod p^\alpha$. Then $[g, z]$ lies in the first layer of P_{i+1}, and not in the second.*

Proof. To say that g lies in the bottom layer of P_i is to say that the g_i are all equal, and this clearly implies that $[g, y]$ defines an element in the augmentation ideal of A, but not in the square of the augmentation ideal, since $i \not\equiv 0 \bmod p^\alpha$. Note that, if $i \equiv 0 \bmod p^\alpha$ then g is congruent, modulo scalars, to an element of P_{i+1}. q.e.d.

(VII.6) Lemma. *Let g lie in the next to bottom layer of P_i, but not in the bottom layer, where $i \equiv 0 \bmod p^\alpha$. Then $[g, y]$ lies in the zero-th layer of P_{i+1}, and not in the first layer.*

Proof. The image of g under $\theta_i\phi$ is $c(1 + 2z + 3z^2 + \cdots + p^\alpha z^{p^\alpha - 1})$ for some non-zero constant c, which we may take to be 1. Then we see at once that the image of $[g, y]$ modulo P_{i+2} under $\theta_{i+1}\phi$ is 1, as required. q.e.d.

To complete our analysis of the lower central series of G, we need to find how low down the series the top layers of the P_i come.

(VII.7) Lemma. *Let $g \in P_i$, where $p^\alpha \not| i$, and let $\nu(i+1) = \beta$, where ν denotes the p-adic valuation of $\mathbb{Z}$. Then $[g, y]$ lies in the top layer of P_{i+1}, and not in the second layer, if g lies in the $p^\alpha - p^\beta - 1$-th layer of P_i/P_{i+1}, and no lower.*

Proof. Let g lie in the $p^\alpha - t$-th layer of G, but no lower. Then we may take $g\theta_i$ to be $(b_1, b_2, \ldots, b_{p^\alpha})$, where $b_j = (-1)^j \binom{p^\alpha - t}{j} = \binom{j+t-1}{t-1}$. Then $[g, y]\theta_{i+1} = (c_1, c_2, \ldots, c_{p^\alpha})$, where $c_1 = -b_2$, and $c_{p^\alpha - i+1} = b_{p^\alpha - i+1}$. So we need to find the least t for which $b_2 \neq b_{p^\alpha - i+1}$.
If $t = 1$ then $b_j = b_k$ for all j, k. So $t > 1$.
If $t = 2$ then $b_j = j + 1$ for all j, so $t = 2$ unless $i \equiv -1 \bmod p$.
Now assume that $i \equiv 1 \bmod p$. More precisely, let $\nu(i+1) = \beta$. If $m = \sum_j m_j p^j$, and $n = \sum_j n_j p^j$, where $0 \leq m_j, n_j \leq p-1$ for all j then

$$\binom{m}{n} \equiv \prod_j \binom{m_j}{n_j} \bmod p.$$

Using this formula, if $p^\alpha - k = 1 + k_\beta p^\beta + k_{\beta+1} + \cdots$, and $t = t_0 + t_1 p + \cdots$, where $0 \leq k_j, t_j \leq p-1$ for all j, then $b_2 \equiv \binom{t+1}{2} \bmod p$, and

$$b_{p^\alpha - k+1} \equiv \binom{t_0 + 1 + t_1 p + \cdots + t_{\beta-1}p^{\beta-1} + (t_\beta + k\beta)p^\beta + \cdots}{t_0 + t_1 p + \cdots + t_{\beta-1}p^{\beta-1} + t_\beta p^\beta + \cdots} \bmod p.$$

Assume now that p is odd. The above expression for $b_{p^\alpha - k+1}$ is not in the form that allows us to apply our formula in all cases. For example, if $t_0 = 0$ or $t_0 = p-1$. But in these cases b_2 and $b_{p^\alpha - k+1}$ are both zero modulo p. It is now easy to see that the least value of t that makes these two terms not equal modulo p is $p^\beta + 1$, as claimed. For $p = 2$ the argument is essentially the same. q.e.d.

We can now read off the structure of the lower central series of G.

(VII.8) Lemma. *Let G_i denote G/P_i, where $i > 1$, and let p be odd. Let c_i be the nilpotency class of G_i. The last term of the lower central series of G_i is of order p, and corresponds to the bottom layer of P_{i-1}/P_i, and $c_{i+1} = c_i + p^\alpha - 1$. The last $p^\alpha - 1$ terms of the lower central series of G_i are of order p. For $j \leq c_i$ the j-th lower central factors of G_i and G_{i+1} are equal, except that the $c_i - p^\beta + 1$-st term of the lower central series of G_{i+1} is the direct product of the corresponding lower central factor for G_i with a p-cycle, where $\beta = \nu(i+2)$.*

(VII.9) Theorem. *The lower central series of G has the following structure if p is odd. $G/\gamma_2(G) \cong C_p \times C_{p^\alpha}$, $\gamma_i(G)/\gamma_{i+1}(G) \cong C_p \times C_p$ if $i = r(p^\alpha - 1) - p^\beta + 2$ for some $r \in \mathbb{N}$ such that $\nu(r+1) = \beta < \alpha$. Otherwise $\gamma_i(G)/\gamma_{i+1}(G) \cong C_p$. In particular the average width is*

$$w_a(G) = \lim_{n \to \infty} \log_p(|G : \gamma_n(G)|)/n = 1 + 1/p^\alpha.$$

Proof. The claims follow from Lemma (VII.8). q.e.d.

If $p = 2$ things are slightly different. Assume in this case that $\alpha \geq 2$. Then, since one has to factor out the centre in characteristic zero, $P_{p^\alpha}/P_{p^\alpha+1}$ is of dimension $p^\alpha - 2$, but is not trivial as $\alpha \geq 2$. Otherwise the same analysis holds as above.

(VII.10) Theorem. *The lower central series of G has the following structure if $p = 2$. $G/\gamma_2(G) \cong C_2 \times C_{2^\alpha}$, $\gamma_i(G)/\gamma_{i+1}(G) \cong C_2 \times C_2$ if $i = r(2^\alpha - 1) - 2^\beta + 2$ for some $r \in \mathbb{N}$ such that $\nu(r+1) = \beta < \alpha$ and, in characteristic zero, $r \neq 2^\alpha$, that is, $i \neq 2^\alpha(2^\alpha - 1) + 1$. Otherwise $\gamma_i(G)/\gamma_{i+1}(G) \cong C_2$. In particular the average width is*

$$w_a(G) = \lim_{n\to\infty} \log_2(|G : \gamma_n(G)|)/n = 1 + 1/2^\alpha.$$

VIII Algorithms on fields

a) Arithmetic in $\mathcal{O}$

We have four sets of algorithms, doing fast or slow arithmetic in characteristic 0 or p in a local ring $\mathcal{O}$ modulo some appropriate power of the (unique) maximal ideal with finite residue class field F. We can only use fast arithmetic if the ring is small enough for us to be able to pack each element into a machine word. In characteristic 0 we also need to be able to construct certain tables, which imposes a stricter condition on the size of the ring. Most examples constructed for these notes were computed using fast arithmetic taking up to 20 seconds; but some required higher accuracy, and needed to be done in slow arithmetic, taking about 8 minutes (on a HP 9000/730). The calculations needing slow arithmetic were those where the field was of relatively high degree over $\mathbb{Q}_p$. Arithmetic in the residue class field F is done in the standard way. That is to say, a primitive element a of F is found, and elements can be stored in additive form as polynomials of degree less than f in a, where $F = GF(p^f)$, or in multiplicative form as a power of a (or zero). Lookup tables are constructed to convert between these representations. Of course the additive form is packed into an integer. We also construct a table that gives, for each i, the integer j such that $a^i + 1 = a^j$ (with suitable conventions for the zero element of the field) and add by using this formula. The characteristic p case is entirely straightforward. Calculating in $F[[t]]/(t^m)$, in slow arithmetic, we store ring elements as arrays of length m, and we simply add, multiply, and invert units, using the naive algorithms. In the fast case, we encode $\sum c_i t^i$ as $\sum c_i q^i$ where the c_i lie in the range $0 \leq c_i < q$ with $q = p^f$. It is a triviality to encode the algorithms for performing these operations without unpacking the elements into arrays. So we perform addition in $O(m)$ integer operations, and multiplication in $O(m^2)$ operations.
Slow arithmetic in characteristic 0 is more delicate. The basic idea is to regard an element of $\mathcal{O}$ as a polynomial over $\mathbb{Z}_p$ of degree $f-1$ in a and degree $e-1$ in π. So we store elements as double arrays of integers, where the entries are taken modulo some power of p. One problem is that we wish to compute modulo π^m for some value of m, and this ideal need not be of the form $p^k\mathcal{O}$; so we work modulo p^k for the least k such that $\pi^m \leq p^k\mathcal{O}$, and then have a test for equality of ring elements that decides whether the value of the difference is at least m. A technical problem arises in that, in order to obtain answers that are correct modulo π^m, we sometimes need, when considering the action of automorphisms, to work to a higher degree of accuracy.

A more interesting problem arises with the element a. This is now a representative in $\mathcal{O}$ of a primitive element of F, and is required to be a precise root of unity. While this is not needed for us to be able to perform slow arithmetic, it is needed for various reasons of efficiency. We start from the minimum polynomial $f_0(x)$ of a modulo π, which is, in effect, the minimum polynomial of the primitive element used to define F. This polynomial is lifted to a polynomial $f_1(x)$ over $\mathcal{O}$ mod π^m by interpreting the coefficients in ($\mathbb{Z}$ mod p) as elements of ($\mathbb{Z}$ mod p^k). Let a_0 be a root of this polynomial. We then compute a as $\lim_i a_0^{p^i}$. This gives an exact root of unity, congruent to a_0 modulo π. We then compute the minimal polynomial $f_2(x)$ of a which is an irreducible factor of the cyclotomic polynomial, and use $f_2(x)$ instead of $f_1(x)$ for further calculations in our ring.

b) Calculating automorphisms in characteristic 0

Any automorphism of the ring $\mathcal{O}$ will take a to a^{p^i} where $0 \leq i < f$, and π to some zero, in $\mathcal{O}$, of the image of the Eisenstein polynomial under the automorphism of the inertia field T that takes a to a^i. The basic problem then is to find all zeros of a given polynomial over $\mathcal{O}$. By Hensel's lemma, a sufficiently accurate approximation to a zero does lift to a unique precise zero, so we are justified in working in $\mathcal{O}/\pi^k\mathcal{O}$ provided that k is big enough to ensure that the Hensel condition is satisfied, and that the accuracy is sufficient for our other calculations. Since the Hensel condition is liable to be the stronger of the two, we may have to compute the automorphisms using slow arithmetic even when our other calculations can be done using fast arithmetic. In practice, we always compute the automorphism using slow arithmetic, as this is fast enough for this purpose. If our matrices have entries in $\mathcal{O}/\pi^m\mathcal{O}$, we need to take k to be strictly greater than m, since we need on occasion to divide by π, and this reduces the accuracy of our approximation by 1. We now work modulo successive powers of π, finding all liftings of zeros accurate to one power of π to zeros accurate to the next power. Also for this procedure one might need a larger quotient of the ring than $\mathcal{O}/\pi^m\mathcal{O}$ to ensure that the approximation of a zero coincides with a zero modulo π^m. The size of the ring depends on the value of the derivative of the Eisenstein polynomial. This naive approach is quite fast enough for polynomials of the small degrees needed.
Each of these zeros, together with the corresponding image of a, defines an automorphism of $\mathcal{O}$, and we calculate the multiplication table, and store this and other basic information about the automorphisms.

c) The group of units of $\mathcal{O}$

If $\mathcal{O}$ is of characteristic p, the group of units of $\mathcal{O}$ is easy to compute. In this case $\mathcal{O}$ is $\mathbb{F}_q[[t]]$ where $q = p^f$, and the group of units is a free $\mathbb{Z}_p$-module with topological basis $\{1 + t^i\}$ where i runs over all natural numbers prime to p.
We now consider the more interesting case of characteristic 0. There is a natural homomorphism of $\mathcal{O}$ onto the residue class field $\mathbb{F}_q$, and this homomorphism induces a split homomorphism of the group U of units of $\mathcal{O}$ onto the cyclic group of order $q-1$. The kernel of this homomorphism is the group U_1 of 1-units of $\mathcal{O}$. As it is well known, U_1 is the direct product of a free $\mathbb{Z}_p$-module of rank $n = ef$ and a finite cyclic p-group occurring from roots of unity. The finite part, which may be trivial, is computed as follows. Let $1 + x$ be of order p, where x is of positive value. Then $(1+x)^p = 1 + px + \cdots + x^p$, and for this to be 1, we need $\nu(px) = \nu(x^p)$, and hence, if the valuation is chosen so that $\nu(\pi) = 1$, we need $e + \nu(x) = p\nu(x)$, and hence e must be a multiple of $p-1$. Taking this calculation to a greater degree of accuracy, let $x = b\pi^k$ modulo higher terms and b some unit. Then computing $(1 + x)^p = 1$

gives $pb\pi^k = b^p\pi^{kp}$ modulo higher terms. So if $p = c\pi^e$ modulo higher terms, we have $bc\pi^{e+k} = b^p\pi^{kp}$, so c must be a $p-1^{\text{th}}$ power. If these conditions (which are vacuous for $p = 2$) are satisfied, it follows from Hensel's lemma that x as required will exist, and will have value $(e-1)/p$. If now $e-1$ is a multiple of p^t for some $t > 1$, then x will have a $p^{t-1^{\text{th}}}$ root, and so the torsion part of U_1 will have order p^t.

A basis for a complement to the torsion part of U_1 can be constructed as follows. If U_1 is torsion-free we take the set of elements of the form $1 + a^i\pi^j$, where $0 \le i \le f$ and $1 \le j < pe/(p-1)$ and j is prime to p. If U_1 is not torsion-free, we make two changes. First we add a new generator $1 + a^\ell\pi^{pe/(p-1)}$ where a^ℓ is not in the image of the endomorphism μ of the additive group of $\mathbb{F}_q$ defined by $x \mapsto cx + x^p$. Secondly we remove a generator $1 + a^i\pi^j$ where $j = e/(p-1)p^\nu$ if p^ν is the highest power of p that divides e, and i is the least integer such that $a^{ip^\nu}\mu$ is an $\mathbb{F}_p$-linear combination of the $a^{up^\nu}\mu$ for $u < i$. (Here we regard a as an element of $\mathbb{F}_q$.)
The above facts are easy to verify. For details see [Iws 86] or [Has 49].

d) Fast arithmetic in characteristic 0

This is done using logarithm tables. Suppose that U_1 is torsion free. It is clear from the above that every element of $\mathcal{O}/\pi^m\mathcal{O}$ can be written uniquely in the form $\pi^\lambda a^k \prod(1+a^i\pi^j)^{u_{i,j}}$, where i and j are chosen as above so that the terms in the product are our chosen free generators for U_1, and the exponents take values bounded by easily computed functions of m. Call this the *multiplicative* form of the element. Similarly, any element can be written uniquely as a $\mathbb{Z}$-linear combination of $\{a^i\pi^j\}$ for $0 \le i < f$ and $0 \le j < \min(m, e)$, where the coefficients are non-negative, and bounded by a suitable function of m. Call this the *additive* form of the element. Now it is easy to add elements in additive form, and to multiply elements in multiplicative form. So when we are going to use fast arithmetic, we first go through all elements in multiplicative form, and compute, and store, their additive form, simultaneously constructing the inverse table. The problem is to construct these tables quickly. This is done by using a number of elementary tricks, for example $x(1 + a^i\pi) = x(1 + a^{i-1}\pi)a - xa + x$. Now addition and multiplication by a are fast operations on elements in additive form, so if we need to compute an element y of the form $a^k\pi^\lambda \prod(1+a^i\pi^j)^{u_{i,j}}$, and we have already computed $a^k\pi^\lambda \prod(1 + a^i\pi^j)^{v_{i,j}}$ both in the case $v_{i_0,j_0} = u_{i_0,j_0} - 1$, $v_{i_0-1,j_0} = u_{i_0-1,j_0} + 1$ for some fixed i_0 and j_0, and u and v agree everywhere else, to give an element z, and also in the case $v_{i_0,j_0} = u_{i_0,j_0} - 1$ for the same i_0 and j_0, with u and v again agreeing elsewhere, to give an element x. Then by the above formula, $y = z - xa + x$, and so is quickly computed.
The case when U_1 has torsion is harder, but similar principles are used. Computing these tables efficiently proved to be a very subtle exercise, and we congratulate Colin Murgatroyd and Matthias Zumbroich for the skill with which they wrote this code.

IX Fields of small degree

a) Extensions of $\mathbb{Q}_2$ of degree 2, 3 and 4

Extensions with a soluble Galois group can be obtained by subsequent abelian extensions which are described by local class field theory. In particular, for every closed subgroup H of finite index in the multiplicative group of a local field k there exists a unique extension K of k in the algebraic closure $\bar{k}$ satisfying $\mathrm{Gal}(K,k) \cong k^*/H$ where $H = N_{K/k}(K)$ is the norm group of K over k, cf. [Iws 86] p. 98-100. The number of different extensions of $\mathbb{Q}_p$ in its algebraic closure $\bar{\mathbb{Q}}_p$ of given inertia degree and ramification index is also known from [Kra 62].
The (unique up to isomorphism) totally unramified extension of degree n over $\mathbb{Q}_p$ is given by a primitive $(p^n-1)^{\text{th}}$ root of unity. In the following an irreducible factor of degree n of the cyclotomic polynomial over $\mathbb{Q}$ is given as a minimal polynomial. The Galois group of this extension is cyclic of order n.
The maximal abelian extension of exponent p of a local field K is denoted by $K_{ab,p}$. Note, $|K_{ab,p} : K| = |K^* : (K^*)^p|$ is finite.

(IX.1) Lemma. *Let k be a normal extension field of finite degree over $\mathbb{Q}_p$. Then $k_{ab,p}$ is a Galois extension.*

Proof. Let $g \in \mathrm{Aut}(\bar{\mathbb{Q}}_p, \mathbb{Q}_p)$. Then $k^g_{ab,p}$ and $k_{ab,p}$ are both extensions of k because k is normal. Moreover the Galois groups over k of the two fields are isomorphic and $\mathrm{Gal}(k_{ab},k) \cong k^*/(k^*)^p$ which is the largest exponent p factor group of k^*. Therefore $k^g_{ab,p} = k_{ab,g}$ and $k_{ab,p}$ is normal. q.e.d.

(IX.2) Lemma. *There are 7 non-isomorphic extensions of $\mathbb{Q}_2$ of degree 2 in its algebraic closure $\bar{\mathbb{Q}}_2$. Minimal polynomials can be chosen as follows.*
Unramified extension: x^2+x+1.
Ramified extension: x^2+2, x^2+6, x^2-2, x^2-6, x^2+2x+2, x^2+2x+6.

Proof. The extensions correspond to the 7 different subgroups of order 2 in $\mathrm{Gal}((\mathbb{Q}_p)_{ab,2}, \mathbb{Q}_2) \cong \mathbb{Q}_2^*/(\mathbb{Q}_2^*)^2 \cong \mathbb{Z}/2\mathbb{Z} \oplus \langle -1 \rangle \oplus \mathbb{Z}_2/2\mathbb{Z}_2 \cong C_2^3$. Constructing the extensions by roots of the equations x^2-a one chooses a set of representatives of the 7 classes of $\mathbb{Q}_2^*/(\mathbb{Q}_2^*)^2$. By suitable substitutions one gets the enumerated polynomials as Eisenstein polynomials or, in the unramified case, as a factor of the cyclotomic polynomial. q.e.d.

(IX.3) Lemma. *There are 2 non-isomorphic extensions of $\mathbb{Q}_2$ of degree 3 in its algebraic closure $\bar{\mathbb{Q}}_2$. Minimal polynomials can be chosen as follows.*
Define $\alpha := \zeta_7 + \zeta_7^2 + \zeta_7^4$ (i.e. a root of x^2+x+2) where ζ_7 is a primitive 7th root of unity.
Unramified extension: $x^3+(1+\alpha)x^2+\alpha x-1$ with Galois group C_3.
Ramified extension: x^3-2 with Galois group S_3.

Proof. Since $\mathbb{Q}_2^*/(\mathbb{Q}_2^*)^3 \cong \mathbb{Z}/3\mathbb{Z}$ there is exactly one normal abelian extension with Galois group C_3 which is clearly the unramified one. To construct an extension with Galois group S_3 one constructs an extension of degree 3 on top of an extension of degree 2 in a non abelian way. There is only one extension of $\mathbb{Q}_2$ with Galois group S_3 namely the above with intermediate field $\mathbb{Q}_2[\zeta_3]$, where ζ_3 is a primitive root of unity, since all other fields of degree 2 (cf. Lemma (IX.2)) yield abelian extensions. q.e.d.

(IX.4) Lemma. *There are 59 non-isomorphic extensions of $\mathbb{Q}_2$ of degree 4 in its algebraic closure $\bar{\mathbb{Q}}_2$. Minimal polynomials can be chosen as follows.*

Define $\alpha := \zeta_{15}^i + \zeta_{15}^{2i} + \zeta_{15}^{4i} + \zeta_{15}^{8i}$ for $i = 1$ or $i = 7$ where ζ_{15} is a primitive 15th root of unity, i.e. α has minimal polynomial $y^2 - y + 4$.
Then, $x^4 + (\alpha - 1)x^3 - 2x^2 - \alpha x + 1$ is a minimal polynomial for the unramified extension with Galois group C_4.

Totally ramified extensions with Galois group C_4:

no	*minimal polynomial*	*factorisation over intermediate field*
33	$x^4 + 4x^2 + 2$	$(x^2 + 2 + \sqrt{2})(x^2 + 2 - \sqrt{2})$
34	$x^4 + 12x^2 + 18$	$(x^2 + 6 + 3\sqrt{2})(x^2 + 6 - 3\sqrt{2})$
35	$x^4 + 20x^2 + 50$	$(x^2 + 10 + 5\sqrt{2})(x^2 + 10 - 5\sqrt{2})$
36	$x^4 + 28x^2 + 98$	$(x^2 + 14 + 7\sqrt{2})(x^2 + 14 - 7\sqrt{2})$
37	$x^4 + 4x^2 + 10$	$(x^2 + 2 + \sqrt{-6})(x^2 + 2 - \sqrt{-6})$
38	$x^4 + 12x^2 + 90$	$(x^2 + 6 + 3\sqrt{-6})(x^2 + 6 - 3\sqrt{-6})$
39	$x^4 + 20x^2 + 250$	$(x^2 + 10 + 5\sqrt{-6})(x^2 + 10 - 5\sqrt{-6})$
40	$x^4 + 28x^2 + 490$	$(x^2 + 14 + 7\sqrt{-6})(x^2 + 14 - 7\sqrt{-6})$

Totally ramified extensions with Galois group V_4:

no	*minimal polynomial*	*roots to be adjoined to intermediate fields*
41	$x^4 + 2x^2 - 4x + 2$	$\sqrt{2}$, $\sqrt{-2}$, $\sqrt{-1}$
42	$x^4 + 14x^2 - 20x + 14$	$\sqrt{2}$, $\sqrt{6}$, $\sqrt{-5}$
43	$x^4 + 4x^3 + 6x^2 + 4x + 10$	$\sqrt{-1}$, $\sqrt{6}$, $\sqrt{-6}$
44	$x^4 + 4x^3 + 2x^2 - 4x + 6$	$\sqrt{-5}$, $\sqrt{10}$, $\sqrt{-2}$

Totally ramified extensions not bi-quadratically constructible:

no	*minimal polynomial*	*Galois group*
45	$x^4 + 2x^3 + 2x^2 + 2$	A_4
46	$x^4 + 2x + 2$	S_4
47	$x^4 + 4x + 2$	S_4
48	$x^4 + 4x^2 + 4x + 2$	S_4

Totally ramified extensions with Galois group D_8:

no	minimal polynomial	factorisation over intermediate field
1	x^4-2	$(x^2+\sqrt{2})(x^2-\sqrt{2})$
2	x^4-18	$(x^2+3\sqrt{2})(x^2-3\sqrt{2})$
3	$x^4+4x^3+8x^2+8x+2$	$(x^2+2x+2+\sqrt{2})(x^2+2x+2-\sqrt{2})$
4	$x^4+4x^3+16x^2+24x-14$	$(x^2+2x+6-5\sqrt{2})(x^2+2x+6+5\sqrt{2})$
5	x^4-6	$(x^2+\sqrt{6})(x^2-\sqrt{6})$
6	x^4-54	$(x^2+3\sqrt{6})(x^2-3\sqrt{6})$
7	x^4+4x^2-2	$(x^2+2-\sqrt{6})(x^2+2+\sqrt{6})$
8	x^4+12x^2-18	$(x^2+6-3\sqrt{6})(x^2+6+3\sqrt{6})$
9	$x^4+4x^3+16x^2+24x-114$	$(x^2+2x+6-5\sqrt{6})$
		$(x^2+2x+6+5\sqrt{6})$
10	$x^4+4x^3+8x^2+8x-2$	$(x^2+2x+2+\sqrt{6})$
		$(x^2+2x+2-\sqrt{6})$
11	x^4+2	$(x^2+\sqrt{-2})(x^2-\sqrt{-2})$
12	x^4+18	$(x^2+3\sqrt{-2})(x^2-3\sqrt{-2})$
13	x^4+4x^2+6	$(x^2+2+\sqrt{-2})(x^2+2-\sqrt{-2})$
14	x^4+12x^2+54	$(x^2+6+3\sqrt{-2})(x^2+6-3\sqrt{-2})$
15	$x^4+4x^3+8x^2+8x+6$	$(x^2+2x+2+\sqrt{-2})$
		$(x^2+2x+2-\sqrt{-2})$
16	$x^4+4x^3+16x^2+24x+86$	$(x^2+2x+6+5\sqrt{-2})$
		$(x^2+2x+6-5\sqrt{-2})$
17	x^4+6	$(x^2+\sqrt{-6})(x^2-\sqrt{-6})$
18	x^4+54	$(x^2+3\sqrt{-6})(x^2-3\sqrt{-6})$
19	$x^4+4x^3+8x^2+8x+10$	$(x^2+2x+2+\sqrt{-6})$
		$(x^2+2x+2-\sqrt{-6})$
20	$x^4+4x^3+16x^2+24x+186$	$(x^2+2x+6+5\sqrt{-6})$
		$(x^2+2x+6-5\sqrt{-6})$
21	x^4-2x^2+2	$(x^2-1+\sqrt{-1})(x^2-1-\sqrt{-1})$
22	x^4-6x^2+18	$(x^2-3-3\sqrt{-1})(x^2-3+3\sqrt{-1})$
23	x^4+6x^2+10	$(x^2+3+\sqrt{-1})(x^2+3-\sqrt{-1})$
24	x^4+18x^2+90	$(x^2+9+3\sqrt{-1})(x^2+9-3\sqrt{-1})$
25	$x^4+4x^3+2x^2-4x+2$	$(x^2+2x-1+\sqrt{-1})$
		$(x^2+2x-1-\sqrt{-1})$
26	x^4-2x^3+2	$(x^2+(-1-\sqrt{-1})x-1+\sqrt{-1})$
		$(x^2+(-1+\sqrt{-1})x-1-\sqrt{-1})$
27	x^4-2x^2+6	$(x^2-1+\sqrt{-5})(x^2-1-\sqrt{-5})$
28	x^4+2x^2+6	$(x^2+1+\sqrt{-5})(x^2+1-\sqrt{-5})$
29	x^4+10x^2+30	$(x^2+5+\sqrt{-5})(x^2+5-\sqrt{-5})$
30	x^4-10x^2+30	$(x^2-5+\sqrt{-5})(x^2-5-\sqrt{-5})$
31	$x^4+4x^3+6x^2+4x+6$	$(x^2+2x+1+\sqrt{-5})$
		$(x^2+2x+1-\sqrt{-5})$
32	$x^4-2x^3+8x^2-32x+46$	$(x^2+(-1-\sqrt{-5})x+1+3\sqrt{-5})$
		$(x^2+(-1+\sqrt{-5})x+1-3\sqrt{-5})$

Ramified extensions ($f = 2$) with Galois group D_8:

no	minimal polynomial	factorisation over intermediate field
49	$x^4 + 12$	$(x^2 + 4\zeta_3 + 2)(x^2 - 4\zeta_3 - 2)$
50	$x^4 + 10x^2 + 28$	$(x^2 + 2\zeta_3 + 6)(x^2 - 2\zeta_3 + 4)$
51	$x^4 + 4x^3 + 6x^2 + 4x + 4$	$(x^2 + 2x + 2 + 2\zeta_3)(x^2 + 2x - 2\zeta_3)$
52	$x^4 + 4x^3 + 2x^2 - 4x + 4$	$(x^2 + 2x + 2\zeta_3)(x^2 + 2x - 2\zeta_3 - 2)$

Ramified ($f = 2$) extensions with Galois group C_4:

no	minimal polynomial	factorisation over intermediate field
53	$x^4 - 4x^2 + 52$	$(x^2 - 8\zeta_3 - 6)(x^2 + 8\zeta_3 + 2)$
54	$x^4 + 4x^3 + 4x^2 + 12$	$(x^2 + 2x - 4\zeta_3 - 2)(x^2 + 2x + 4\zeta_3 + 2)$
55	$x^4 + 4x^2 + 52$	$(x^2 + 8\zeta_3 + 6)(x^2 - 8\zeta_3 - 2)$

Ramified ($f = 2$) extensions with Galois group V_4:

no	minimal polynomial	roots to be adjoint to intermediate fields
56	$x^4 - 2x^2 + 4$	$\sqrt{-3}, \sqrt{-2}, \sqrt{6}$
57	$x^4 - 6x^2 + 36$	$\sqrt{-3}, \sqrt{2}, \sqrt{-6}$
58	$x^4 - 2x^3 + 2x^2 - 4x + 4$	$\sqrt{-3}, \sqrt{-1}, \sqrt{3}$

Proof. Those extensions which have a quadratic intermediate field K can be constructed by taking quadratic extensions of quadratic extensions each step according to Lemma (IX.2). The extensions might not be normal. Checking equality of the so constructed extensions is an easy task.

The missing extensions have S_4 or A_4 as Galois groups. First consider those with A_4. According to the unique chief series of A_4 one has to construct first the unique (up to isomorphism) unramified extension by $\theta = \zeta_7 + \zeta_7^{-1}$ which has $C_3 = \langle\sigma\rangle$ as Galois group. Now one constructs a Galois extension K of $k = \mathbb{Q}_2[\theta]$ with $\mathrm{Gal}(K, \mathbb{Q}_2) \cong A_4$ such that the Galois group $\mathrm{Gal}(K, k)$ is V_4. Independent generators of the $k^*/(k^*)^2$ are 1, 2, 5, 7, $1 + \theta$ and $1 + \theta^\sigma$. Since $k^*/(k^*)^2$ obviously splits as a $\mathbb{F}_2 C_3$-module into a direct sum of 4 simple modules with trivial action of σ and one 2-dimensional module. Adjoining $\sqrt{-1-\theta}$ and $\sqrt{1-\theta^2}$ to k, the fixed field of σ is the required extension of degree 4 over $\mathbb{Q}_2$.

To construct the fields with S_4 as Galois group, one extends $k = \mathbb{Q}_2[\sqrt[3]{2}, \zeta_3]$ (cf. Lemma (IX.3)) to K with $G := \mathrm{Gal}(K, \mathbb{Q}_2) \cong S_3$ such that $\mathrm{Gal}(K, k) = V_4$. Let $N = k^*/(k^*)^2 \cong C_2^8 \cong \mathrm{Gal}(k_{ab,2}, \mathbb{Q}_2)$ and σ be an element of order 3 in S_3. Define $Z := Fix_N(\sigma) \cong C_2^4$. Therefore N, under the action of an element of order 3 in G, splits into $N = Z \oplus M$, where M consists of 2 copies of the 2-dimensional irreducible $\mathbb{F}_2 C_3$-module. This structure is preserved when the representations of C_3 on M is extended to S_3. Since $M \cong N/Z$ there are 3 normal subgroups in this factor with the pre-images M_1, M_2, M_3 such that $NG/M_i \cong S_4$. Therefore there are 3 extension fields of $\mathbb{Q}_2$ with Galois group S_4. The fixed fields under G are the required fields. q.e.d.

b) Extensions of $\mathbb{Q}_3$ of degree 2, 3 and 4

(IX.5) Lemma.

(i) There are 3 non-isomorphic extensions of $\mathbb{Q}_3$ of degree 2 in its algebraic closure $\bar{\mathbb{Q}}_3$. Minimal polynomials can be chosen as follows.
Define α to be $\zeta_8^i + \zeta_8^{3i}$ where ζ_8 is a primitive 8th root of unity and $i = 1$ or 5 (i.e. α is a root of $x^2 + 2$).
Unramified extension: $x^2 + \alpha x - 1$.
Ramified extensions: $x^2 + 3$, $x^2 - 3$.

(ii) There are 10 non-isomorphic extensions of $\mathbb{Q}_3$ of degree 3 in its algebraic closure $\bar{\mathbb{Q}}_3$. Minimal polynomials can be chosen as follows.
Define α to be a root of $y^4 + y^3 + 2y^2 - 4y + 3$ i.e. any of $\zeta_{13}^i + \zeta_{13}^{3i} + \zeta_{13}^{9i}$ for $i = 1, 2, 4, 7$ where ζ_{13} is a primitive 13th root of unity.
Unramified extension: $x^3 + \alpha x^2 + (1 - \frac{2}{3}\alpha - \frac{1}{3}\alpha^3)x + 1$.

Ramified extensions:

minimal polynomial	*Galois group*	*intermediate field of degree 2*
$x^3 + 3x^2 + 3$	S_3	$\mathbb{Q}_3[\zeta_8]$
$x^3 + 3$	S_3	$\mathbb{Q}_3[\sqrt{-3}]$
$x^3 + 12$	S_3	$\mathbb{Q}_3[\sqrt{-3}]$
$x^3 - 6$	S_3	$\mathbb{Q}_3[\sqrt{-3}]$
$x^3 + 3x^2 + 3x + 3$	S_3	$\mathbb{Q}_3[\sqrt{-3}]$
$x^3 + 3x^2 - 3x + 3$	S_3	$\mathbb{Q}_3[\sqrt{3}]$
$x^3 + 3x^2 - 3$	C_3	
$x^3 + 3x^2 - 12$	C_3	
$x^3 + 3x^2 + 6$	C_3	

(iii) There are 5 non-isomorphic extensions of $\mathbb{Q}_3$ of degree 4 in its algebraic closure $\bar{\mathbb{Q}}_3$. Minimal polynomials can be chosen as follows.
Define $\alpha := \zeta_{80} + \zeta_{80}^3 + \zeta_{80}^9 + \zeta_{80}^{27}$ to be a primitive 80th root of unity i.e. α a root of $x^8 + 60x^4 + 200x^2 + 225$.
Unramified extension:
$x^4 - \alpha x^3 + 135^{-1}(\alpha^6 + \alpha^4 + 115\alpha^2 + 180)x^2 + 135^{-1}(-\alpha^7 - \alpha^5 - 70\alpha^3 - 225\alpha)x - 1.$
Define ζ_8 to be a primitive 8th root of unity.
Ramified extension over the unramified field $\mathbb{Q}_3(\zeta_8)$ of degree 2:
$x^2 + 3$ (Galois group V_4),
$x^2 + 3\zeta_8$ (Galois group C_4).
Totally ramified extension:
$x^4 + 3$ (Galois group D_8),
$x^4 - 3$ (Galois group D_8).

Proof. (i) Straightforward.
(ii) There are 4 normal extensions with Galois group C_3 corresponding to the different subgroups isomorphic to C_3 in $\mathbb{Q}_3^*/(\mathbb{Q}_3^*)^3$. To construct the other extensions of degree 3 one has to consider normal extensions with Galois group S_3 and the required fields are those corresponding to subgroups of index 3. On top of any of the 3 extensions k

in (i) one constructs extensions of degree 2 which have non-abelian Galois groups over $\mathbb{Q}_3$. This is only possible in one way when $k = \mathbb{Q}_3[\sqrt{3}]$ or $k = \mathbb{Q}_3[\zeta_8]$. For $k = \mathbb{Q}_p[\sqrt{-3}]$ one has to consider the operation of C_2 on $\mathrm{Gal}(k_{ab,3}, k) \cong k^*/(k^*)^3 \cong C_3 \times C_3 \times C_3 \times C_3$ which splits into a two dimensional module on which C_2 acts trivially and a two dimensional module on which C_2 acts non-trivially. Hence one gets 4 different extensions corresponding to the 1-dimensional non-trivial submodules.
(iii) The given polynomials are those which have a quadratic intermediate field and therefore can be constructed by taking quadratic extensions of quadratic extensions, cf. Lemma (IX.4). Assume there are extensions K of $\mathbb{Q}_3$ with Galois group G isomorphic to A_4 or S_4. Then G contains certain normal subgroups $G_1, \ldots, G_n$ for some $n \in \mathbb{N}$ with $G \geq G_0 \geq G_1 \geq \cdots \geq G_n$, called ramification groups in the upper numbering cf. [Iws 86] p. 33-34. These subgroups satisfy $|G : G_0| = f$, $|G_0 : G_1| \mid (p^f - 1)$ and $|G_1 : \langle 1 \rangle| = p^k$ for some $k \in \mathbb{N}$ and f the inertia degree of K. Comparing these conditions with the (unique) chief series of G yields a contradiction. q.e.d.

X Algorithm for finding a filtration and the obliquity

a) The BASIS algorithm

Let P be a finite p-group, with a composition series

$$P = P_1 > P_2 > \cdots > P_{n+1} = \langle 1 \rangle.(*)$$

For $1 \leq i \leq n$, let $a_i \in P_i - P_{i+1}$ be fixed. So $\{a_i\}$ forms a *basis* for P. It is useful to have $(*)$ as a chief series for P, that is with $P_i \lhd P$ for all i, but this condition will not always be satisfied.

If $g \in P$, then g can be written uniquely in the *normal form* $g = a_1^{u_1} \ldots a_n^{u_n}$, where $0 \leq u_i < p$. Assume that we have an algorithm for multiplying and inverting elements of P, and an algorithm for writing elements of P in normal form. Then we can compute in normal form the p-th powers of the basis elements, and the commutator of each pair of basis elements, thus producing a *PC-presentation* of P. A very wide range of algorithms are available, and have been implemented in the computer systems GAP [GAP 94] and MAGMA [MAG 95], for computing with p-groups (and soluble groups) defined in terms of PC-presentations. These algorithms, in most cases, simply require one to be able to multiply and invert elements, and to express the result in normal form, which we could do more efficiently without use of the PC-presentation. But constructing a PC-presentation for the groups in question, and sending the results to a file, enables us to analyse the groups we have constructed in a form that is independent of their provenance, and enables us to use the above packages.

The basic algorithm is performed in the following context. We have a finite p-group G, and a subset X of G. Let $P = \langle X \rangle$. We aim to construct a basis for P. When the algorithm has been described, it will be clear how to write any element in normal form with respect to this basis, and hence how to construct the PC-presentation on this basis.

We assume, of course, that we can multiply and invert in G, and can test for equality. There will be a fixed normal series $G = G_{(1)} > G_{(2)} > \cdots > G_{(wt+1)} = \langle 1 \rangle$ with elementary abelian sections, refined by the series (G_i). If $g \in G_{(i)} - G_{(i+1)}$ we say that g has *weight* i, and define the weight of 1 to be $wt+1$. We assume that we can calculate the weight of any element of G, and if $g \neq 1$ has weight i, we assume we can compute the image of g in $G_{(i)}/G_{(i+1)}$ as a vector with respect to some fixed basis of $G_{(i)}/G_{(i+1)}$.

The algorithm to construct a basis for P goes as follows.

```
BASIS(X, B, wt)
/* wt is as above; X is a finite sequence of elements of G.
B is set to a sequence of elements of G that forms a basis
for P = ⟨X⟩.*/
{ t := length(X);
  B := ∅;
  for w := 1 to wt do
     {g := FIRST(X);
     repeat
          if weight(g) = w;
              {FILTER(w, g, B);
                    if weight(g) = w
                        {B := B ∪ {g};
                        X := X concatenate seq(g^{X_1}, ..., g^{X_t}, g^p);
                        }
                    else X := X concatenate(g);
              }
          }
     until NEXT(X, g) = FALSE;
}
```

The procedure FILTER does the following: it takes as input an integer w, an element g of weight w, and a set B that is part of the basis being constructed. If the image of g in $G_{(w)}/G_{(w+1)}$ is linearly dependent on the images in $G_{(w)}/G_{(w+1)}$ of the elements of B of weight w, then g is multiplied by elements of B of weight w to produce a new g of greater weight. This modified element is again called g. Otherwise, the procedure does nothing. FIRST(X) returns the first element of X, and NEXT(X, g) replaces g by the next element of X, or returns FALSE if g is the last element of X.

b) Split and Non-Split groups

It is unfortunate that an algorithm that is so easy to describe should have required a C program of over 12000 lines.
It should be admitted that the program as implemented does not quite fit the above description. We compute the PC-presentation at the same time as we compute the basis, and this enables us to use the power-commutator presentation to drive the construction of new elements of B. However, this is a minor variation. The main difficulty lies in computing with elements of G.

The group G will have a somewhat more complicated structure than described above. It will in fact have a normal series, of length at most four whose factors will have the structure described; in particular, each factor will have a weight function.

The weight functions of these factors define, in a natural way, a filtration of G, and the subgroups of G that arise in the filtration are required to be normal in G. It is this normal series that is used in the procedure BASIS described above.

We now distinguish two cases. In the simplest, or 'split' case, each extension splits, and we can express G as $G[3] \rtimes G[2] \rtimes G[1] \rtimes G[0]$, where each $G[i]$ is embedded as a subgroup of G. An element of G that lies in a $G[i]$ will be called *simple*. It turns out that we can choose the embedding of each $G[i]$ as a subgroup of G in such a way that the commutator of any two simple elements is again simple. This implies that, for any $i < j$, either $G[j]$, as a subgroup of G, is normalised by $G[i]$, or that $G[j]$, as a section of G, is centralised by $G[i]$. As a consequence of the splittings and the commutator condition, if the generating set X of P consists entirely of simple elements, then only simple elements arise in the calculation. This will be assumed in the split case; so it is sufficient to write code to perform the basic group-theoretic operations in each $G[i]$, to compute the commutator of two simple elements, and to compute the weight of a simple element. More precisely, the filter procedure requires us to be able to compute the image of an element of weight k in $G[i]_{(k)}/G[i]_{(k+1)}$.

c) The groups $G[i]$

The group $G[3]$, which is in general the largest of the $G[i]$, and the one in which most computation is performed, is defined as follows. Take a local field K with ring of integers $\mathcal{O}$. Then the ring $\mathcal{O}_d$ of $d \times d$ matrices over $\mathcal{O}$ has an ideal I (which is the radical of the standard minimal hereditary order in $\mathcal{O}_d$) consisting of those matrices in $\mathcal{O}_d$ whose entries on and below the main diagonal lie in the maximal ideal of $\mathcal{O}$. Since I is pro-nilpotent, the group $H = H(d, K)$ of matrices of the form $I_d + M$, where I_d is the identity $d \times d$ matrix, and M runs through the elements of I, is a pro-p-group.

Defining $H_{(i)}$ to be the subgroup of H consisting of elements of the form $I_d + M$ where M lies in I^i defines a filtration for H by normal subgroups with elementary abelian sections of rank $d \cdot f$, where f is the degree of the inertia field; that is to say, the degree, over $\mathbb{Q}_p$, of the maximal unramified extension of $\mathbb{Q}_p$ that lies in K. We take $G[3]$ to be a subgroup of the quotient of H by $H_{(w+1)}$ for some $w > 0$, or, more often, of the quotient of H by the product of $H_{(w+1)}$ and the group of scalar matrices in H. To distinguish between these cases, we refer to the former as the *linear* and the latter as the *projective* case.

The filtration $\{G[3]_{(t)}\}$ of $G[3]$ induced by $\{H_{(t)}\}$ is the filtration used in the BASIS algorithm.

Since we work modulo $G[3]_{(w+1)}$, where w is fixed, it is sufficient to take the coefficients of our matrices to lie in the quotient of $\mathcal{O}$ by a suitable power of I. However, matrices that represent the same element of $G[3]$ may still have different coefficients modulo this power of I. Before considering the details, let us agree to follow the inevitable practice of the program, and regard a matrix, with coefficients in a suitable quotient of $\mathcal{O}$, as an element of $G[3]$. This abuse of notation is not without dangers at the programming level.

Suppose that we are in the linear case.

To prove that a matrix $M = (m_{ij})$ lies in $G[3]_{(t)}$ one has simply to check that $\nu(m_{ij}) \geq 1 + (t+i-j-1)/d$ for all $i \neq j$, and that $\nu(m_{ii}-1) \geq 1+(t-1)/d$ for all i, as can be easily seen; so, given a basis for the residue class field of K over its prime field, it is easy to compute the image of M modulo $G[3]_{(t+1)}$. We shall illustrate the method by an example. Suppose then that

$$M = \begin{pmatrix} 1+\pi a_{11} & a_{12} & a_{13} \\ \pi a_{21} & 1+\pi a_{22} & a_{23} \\ \pi a_{31} & \pi a_{32} & 1+\pi a_{33} \end{pmatrix}$$

is an element of $G[3]$. Here, and throughout this section, π denotes a fixed uniformising element of K; that is to say, an element of $\mathcal{O}$ of value 1. To compute the image of M in $G[3]/G[3]_{(2)}$, form the image of (a_{12}, a_{23}, a_{31}) in F^3, where F is the residue class field. If this image is zero, to compute the image of M in $G[3]_{(2)}/G[3]_{(3)}$, form the image of (a_{13}, a_{21}, a_{32}) in F^3. If this image is zero, to compute the image of M in $G[3]_{(3)}/G[3]_{(4)}$, form the image of (a_{11}, a_{22}, a_{33}) in F^3. If this is zero, to compute the image of M in $G[3]_{(4)}/G[3]_{(5)}$, form the image of $(\pi^{-1}a_{12}, \pi^{-1}a_{23}, \pi^{-1}a_{32})$ in F^3, and so on.
If we are in the projective case, everything is the same except on the diagonal. In this case, in order to check that M lies in $G[3]_{(t)}$, we have to see that

$$\nu(a_{ii} - a_{11}) \geq 1 + (t-1)/d \text{ for all } i > 1.$$

Similarly, to determine the image of M as above in $G[3]_{(3)}/G[3]_{(4)}$, which is now of dimension 2 over F, form the image of $(a_{22}-a_{11}, a_{33}-a_{11})$.

It remains to explain how arithmetic is performed in $\mathcal{O}$, or rather in quotients of this ring. This has been discussed in Chapter VIII.

We now discuss the group $G[2]$. The group H is a maximal pro-p-subgroup of $GL(d, \mathcal{O})$, and can be described geometrically as follows. There is a chain of lattices $L = L_0 > L_1 > L_2 > \cdots$ where L is the natural module for $GL(d, \mathcal{O})$, where $L_{i+d} = \pi L_i$ for all i, and H is the group that stabilises this chain, and centralises the quotients. In fact we have taken $L_2 = \langle \pi e_1, e_2, \ldots, e_d \rangle$, and $L_3 = \langle \pi e_1, \pi e_2, \ldots, e_d \rangle$, *etc.* Working over K we can extend this chain to negative exponents by defining $L_{i-d} = \pi^{-1} L_i$. Then the normaliser of H in $GL(d, K)$ is the subgroup that normalises this extended chain. If we work linearly, the quotient of this normaliser by H is a split extension $N_2 \rtimes N_1$, where N_2 is the subgroup of the normaliser that normalises each L_i, and N_1 is generated by an element that maps L_i onto L_{i+1} for all i. Thus N_2 is the direct product of d copies of F^*, and we can take N_1 to be generated by the matrix whose non-zero elements consist of 1 in each place immediately above the main diagonal, and π_0 in the bottom left corner, where π_0 is a uniformising element. If we work projectively, N_1 becomes of order d, as its d-th power is $\pi_0 I_d$. Thus, if p divides d, we can extend H, and hence $G[3]$, by a cyclic group of order p^k, where k is the p-adic value of d. Note that $N_2 \rtimes N_1$ has several Sylow p-subgroups, and that different choices of π_0 mod I^2 will correspond to different Sylow subgroups. Because

of the commutator condition mentioned above (i.e. a commutator of two simple elements is simple), we will need to be careful of our choice of π_0.
Naturally we define $G[2]$ to be this cyclic group of order p^k, or to be trivial if we wish to omit this group. If $G[2]$ is to be non-trivial, we must work projectively for invertibility.
As $G[2]$ is cyclic, the weight function on $G[2]$ is forced, and does not need discussion. To compute the commutator of an element of $G[3]$ with an element of $G[2]$ involves dividing an element of our ring, of positive value, by π_0. As we are only working in a quotient of $\mathcal{O}$, unfortunately π_0 is no longer a unit, so we have to be very careful at this point. Of course the answer mathematically speaking is to compute to a greater accuracy than is needed for the final result, but this causes problems of efficiency.

The group $G[1]$ is a Sylow p-subgroup of the automorphism group of K if K is of characteristic 0, and is a Sylow p-subgroup of the group of automorphisms of K that fix a given uniform element if K is of characteristic p.
The case when K is of characteristic p is the simpler, and we dispose of this case first. The group of automorphisms of K that act as the identity on the residue class field F of K is a so-called Nottingham group. It is known to contain a copy of every finite p-group, and Rachel Camina has recently proved that it even contains a copy of every finitely generated pro-p-group (cf. [Cam 97]). It is, however, difficult to find explicitly an example of any finite subgroup of order greater than p. We have therefore written our code to exclude subgroups of the Nottingham group acting on the matrix entries. Elements of the ring are represented as polynomials in some fixed indeterminate t, and we simply take the p-automorphisms of K that fix t, or we can take $G[1]$ to be trivial. Thus $G[1]$ is cyclic of order p^k where k is the p-adic value of f, and $F = GF(p^f)$, or $k = 0$. Provided that we take the uniform element π_0 in the definition of $G[2]$ to be t, elements of $G[2]$ and $G[1]$ commute, and there is no problem with the action of $G[1]$ on $G[3]$.
If K has characteristic 0, we compute its full automorphism group. The theory behind our calculations is well known; the original reference is [Has 49]. For a more recent reference, see for example [Iws 86]. We supply e, f and the Eisenstein polynomial, where e is the ramification index and f is the degree of inertia of K over $\mathbb{Q}_p$. That is to say, K is obtained first by forming an unramified extension T of $\mathbb{Q}_p$, where T is of degree f over $\mathbb{Q}_p$, and is formed by adjoining the $(p^f-1)^{\text{th}}$ roots of unity to $\mathbb{Q}_p$. Denote a primitive $(p^f-1)^{\text{th}}$ root of unity by a. The automorphism group of T does not act transitively on the set of primitive roots, so the question arises which primitive root has been chosen. The choice is made automatically, and is printed to the screen, the choice being determined by the minimum polynomial of a over $\mathbb{Z}_p$ reduced modulo p. The Eisenstein polynomial is then provided as a monic polynomial of degree e over T, where the coefficients are given as polynomials in a over $\mathbb{Q}_p$. We shall take π to be some fixed zero in K of the given Eisenstein polynomial.
Some details of the calculation of the automorphism group of K have been given in Chapter VIII b). Here we give an outline of the general theory.
The automorphism group Γ of K has a filtration $\Gamma \geq \Gamma_0 \geq \Gamma_1 \geq \cdots$ by normal subgroups, terminating in $\langle 1 \rangle$, where Γ_i is the subgroup that centralises $\mathcal{O}/\pi^{i+1}\mathcal{O}$. It is easy to see that the first two filtration quotients are cyclic of orders dividing f and $p^f - 1$ respectively, and that the others are elementary abelian p-groups.

The first step is to compute Γ, finding the image of a and π under each element, as described later, and to construct the multiplication group of Γ.
The next step is to find a Sylow p-subgroup of Γ. Clearly we may work modulo the normal p-subgroup Γ_1, and restrict ourselves to the subgroup consisting of elements whose image modulo Γ_0 is of p-power order. We now have a meta-cyclic group with a natural action on $\pi\mathcal{O}/(\pi\mathcal{O})^2$. Clearly any p-subgroup must fix a non-zero coset of $(\mathcal{O}\pi)^2$, and it is easy to see that a Sylow p-subgroup is the centraliser of some suitable coset. Thus we can find some $\alpha \in F^*$ such that the subgroup of Γ that acts on $\mathcal{O}/\pi\mathcal{O}$ as field automorphisms of prime power order, and fixes $\alpha\pi \bmod \pi^2$ is a Sylow p-subgroup of Γ. This Sylow subgroup is our $G[1]$.
Having chosen $G[1]$, we need to check the commutator condition. Clearly $G[1]$ normalises $G[3]$, and permutes the Sylow subgroups of the normaliser of H in $PGL(d, K)$. One sees at once that, provided that π_0 is chosen to be congruent to $\alpha\pi$ modulo $(\pi\mathcal{O})^2$, $G[1]$ will centralise $G[2]$ as a section of G, and all is well.
Of course it is possible to choose the Eisenstein polynomial such that α may be taken to be 1.
The filtration we use for $G[1]$ is obtained from the filtration defined above (which is known as the upper numbering) by refining the cyclic quotient Γ/Γ_0 to have factors of order p, and deleting trivial quotients. It should be remarked that Γ/Γ_0 need not centralise Γ_i/Γ_{i+1} for all i, so in general the filtration quotients that we obtain need not be central. Thus the procedure BASIS will not always produce an AG-system that corresponds to a chief series.

Finally if $p = 2$ and $d > 2$, we may take $G[0]$ to be the cyclic group of order 2 that acts on elements of $G[3]$ by the inverse transpose automorphism. It commutes with $G[1]$ and $G[2]$. If $d = 2$ this would simply act as conjugation by an element of the normaliser of H in $GL(d, K)$; this corresponds to the fact that the Dynkin diagram for $A(n)$ has no non-trivial automorphism if $n = 1$, as it consists of a single point.

Unfortunately, we also have to consider non-split examples. It would be possible to deal with these by embedding the groups in larger split extensions, but this would have involved serious calculations to produce the required input. We therefore allow the program to run in non-split mode. In this case we represent an element of G as a linked list of length at most 4, whose entries are non-trivial elements of the $G[i]$, arranged in strictly increasing order of i. This has the advantage that the identity element of G is represented by the null list. We now multiply and divide using collection. This enables us to remove the condition that the product and commutator of two simple elements should be simple. In practice we are only interested in the case in which this can apply to two field automorphisms. So in the non-split case, we have a different action of a field automorphism on a matrix, and the product (or commutator) of two simple elements in $G[1]$ can be the product of an element in $G[3]$ with the corresponding product (or commutator) in $G[2]$. These elements of $G[3]$, which we call cocycles, are computed in advance. So in the non-split case, we have to supply an algorithm for computing these cocycles, and an algorithm for computing the action of $G[1]$ on $G[3]$. There will also be a restriction on $G[3]$; but this will be supplied automatically by the input matrices.

d) Calculating the obliquity

By the algorithm BASIS one can assume to have constructed a PC-presentation for some finite p-group P. Now an algorithm for calculating the obliquity is described. Define $\mathcal{A}_i := \{N \lhd P | N \not\leq \gamma_{i+1}(P)\}$. Recall $\mu_i(P) := (\bigcap_{N \in \mathcal{A}_i} N) \cap \gamma_{i+1}(P)$. The ith obliquity is defined to be $o_i(P) := log_p(|\gamma_{i+1}(P) : \mu_i(P)|)$. In particular $\mathcal{A}_i \subset \mathcal{A}_{i+1}$. Define $\widetilde{\mathcal{A}}_i := \{N \in \mathcal{A}_i | N \subset \gamma_i(P)\}$. Therefore the first observation is that if one has determined $\mu_i(P)$ then

$$\mu_{i+1}(P) = \mu_i(P) \cap (\bigcap_{N \in \widetilde{\mathcal{A}}_{i+1}} N) \cap \gamma_{i+2}(P).$$

Set $\mathcal{M}_i := \{N \lhd P | N \in \widetilde{\mathcal{A}}_i \text{ and } \nexists M \in \widetilde{\mathcal{A}}_i \text{ with } M \subsetneq N\}$. Clearly

$$\mu_{i+1}(P) = \mu_i(P) \cap (\bigcap_{N \in \mathcal{M}_{i+1}} N) \cap \gamma_{i+2}(P) \text{ for } i \in \mathbb{N}.$$

For determining $\mathcal{M}_i$ one uses the following algorithm:

$L :=$ a set of candidates,
$K :=$ some set of normal subgroups,
$max(X)$ a function returning the set of all maximal subgroups of X which are normal in P

```
{
M_i = ∅;
L := max(γ_i(P));
for N∈ L do
    {
    K := max(N);
    if M ⊂ γ_{i+1}(P) for all M ∈ K then
                        M_i := M_i ∪ N;
    else {
                        for M ∈ K do
                        {
                            if M ⊄ γ_{i+1}(P) then
                                L := L ∪ M;
                            K := K − M;
                        }
    }
    L := L − N;
}
```

To prove that the algorithm finds all normal subgroups of $\mathcal{M}_i$ one uses that P is a p-group and therefore for $M \in \mathcal{M}_i$ there exists a chain of normal subgroups such that $\gamma_i(P) = X_1 > X_2 > \ldots > X_s = M$ with $|X_i : X_{i+1}| = p$ for $i = 1, \ldots, s-1$.

The function $max(N)$ returns the maximal normal subgroups of N which are normal in P. This is done by the following method:

The group P acts on $V := N/[N,N]N^p$ which is an $\mathbb{F}_p$-module. One determines the 1-dimensional eigenspaces of V under the action of P. Since P is a p-group one has to consider only the eigenvalue 1. Call these eigenspaces $\Xi_1, \ldots, \Xi_s$. Since the epimorphisms $\pi_i : V \to \Xi_i$ are P-admissible the kernels $Ker(\pi_i)$ are invariant under the action of P. It follows that the set of pre-images $\{Ker(\pi_i)[N,N]N^p | i = 1, \ldots, s\}$ is the set of all maximal subgroups of N which are normal in P.

e) Periodicity of the lower central series and obliquity

Let G be an insoluble $\tilde{p}$-group. By Theorems (IV.6) and (IV.14) the isomorphism types of the factors in the lower central series $\gamma_i(G)/\gamma_{i+1}(G)$ and the sequence o_j of the obliquity repeat periodically for some $i \geq i_0$ and some $j \geq j_0$. To calculate the tables in Chapter XII it is necessary to determine i_0 and j_0 and a c, such that the calculations can be carried out in a finite quotient $\bar{G} = G/\gamma_{c+1}(G)$ and repeat from i_0 or respectively j_0 on. Denote the $\mathbb{Q}_p$-dimension of the corresponding Lie algebra by d. One can check that the groups investigated in Chapter XII are settled with respect to d. Therefore the periodicity of the lower central series will be guaranteed from $\gamma_d(G)$ onward or if $p = 2$ from the largest $\gamma_m(G)$ which is contained in $\gamma_d(G)^2$ onward ($m \leq 2d$). Unfortunately starting the investigation of the lower central series or obliquity from $\gamma_d(G)$, or $\gamma_m(G)$, requires a rather big quotient $\bar{G}$ of the group G. A more detailed analysis, as carried out below, can reduce the size of the quotient and therefore increase the number of groups which we actually can handle.

We can observe that usually the lower central series shows already the pattern from some $i_1 < i_0$. It can be proved by induction, that if one has an $i_1 \in \mathbb{N}$ and there are $r, k \in \mathbb{N}$ such that $\gamma_i(G)^{p^r} = \gamma_{i+k}(G)$ and $\gamma_i(G)$ is powerful for $i = i_1, \ldots, i_1+k-1$ and $|\gamma_{i_1}(G) : \gamma_{i_1}^{p^r}(G)|$ is a multiple of d (all this can be checked in a suitable quotient $\bar{G}$) then these conditions hold for all $i \geq i_1$.

To calculate the obliquity recall $\mu_i(G) := (\bigcap_{M \not\leq \gamma_{i+1}(G), M \triangleleft G} M) \cap \gamma_{i+1}(G)$. The lattice of normal subgroups contained in $\gamma_d(G)$, or $\gamma_d(G)^2$ if $p = 2$, repeats periodically. The sequence (o_j) is periodic from j_0 onward, if all normal subgroups M which are to be considered for calculating $\mu_{j_0}(G)$ are contained in $\gamma_d(G)$ or $\gamma_d(G)^2$.

A careful analysis also might give us a smaller index j_1 from where on the obliquity repeats. Let N be a normal subgroup in G. Let i be the maximal index, such that $N \leq \gamma_i(G)$. Let $s = s(N)$ be minimal, such that $\gamma_{i+s}(G) \leq N$.
To determine the period of the sequence $(|\gamma_{i+1}(G) : \mu_i(G)|)$, and to prove the repetition, one first estimates $\sigma \in \mathbb{N}$ and $j \geq i_1$ such that $max\{s(N) \mid N \triangleleft G, N \leq \gamma_j(G)\} \leq \sigma$ where i_1 is where the period of the lower central series begins, cf. Lemma (X.1). In Lemma (X.2) one finds some $z = z(\sigma, k, j)$ such that all normal subgroups of G contained in $\gamma_z(G)$ are uniform. It then remains to make sure that all normal subgroups which are necessary to calculate $\mu_i(G)$ are contained in $\gamma_z(G)$. This will be the case for all $\mu_i(G)$ with $i \geq \alpha$ where α is determined in Lemma (X.3).

(X.1) Lemma. *Assume that i_1 and the length of the period k are determined as above such that $\gamma_n(G)$ is powerful and $\gamma_n(G)/\gamma_{n+1}(G) \cong \gamma_{n+k}(G)/\gamma_{n+k+1}(G)$ for all*

$n \geq i_1$. Let $\sigma := max\{s(N) | N \lhd G, N \leq \gamma_{i_1}(G), N \not\leq \gamma_{i_1+k}(G)\}$. Let $j = max\{i_1, \sigma + 1\}$ then

$$max\{s(N) | N \lhd G \text{ with } N \leq \gamma_j(G)\} = \sigma.$$

Proof. Claim: The maximal value of σ is reached by normal subgroups generated by one element as a normal subgroup.
One has $\langle x, y\rangle^G \geq \langle x\rangle^G\langle y\rangle^G$. Let $s_1 := s(\langle x\rangle^G)$, $s_2 := s(\langle y\rangle^G)$ und n, l such that $\gamma_{n+s_1}(G) \leq \langle x\rangle^G \leq \gamma_n(G)$ and $\gamma_{l+s_2}(G) \leq \langle y\rangle^G \leq \gamma_l(G)$. Then $\gamma_{max\{n+s_1,l+s_2\}}(G) \leq \langle x\rangle^G\langle y\rangle^G \leq \langle x, y\rangle^G \leq \gamma_{max\{n,l\}}(G)$. It follows that $max\{n + s_1, l + s_2\} - max\{n, l\} \leq max\{s_1, s_2\}$. Hence it holds that $s(\langle x, y\rangle^G) \leq max\{s_1, s_2\}$ and therefore the claim, since all normal subgroups of G are finitely generated.
Choose an $x \in N$ where N is a normal subgroup in G satisfying $N \leq \gamma_n(G)$, for $i \geq j + k$. Then, $n = j + f + kl$ with $0 \leq f < k$. For any $x \in N$ there exits an $\tilde{x} \in \gamma_{j+f}(G)$ such that $\tilde{x}^{p^l} = x$. From the assumptions follows $s(\langle\tilde{x}\rangle^G) \leq \sigma$. The factor group $\langle\tilde{x}\rangle^G/\gamma_{j+f+\sigma+1}(G)$ is abelian since $\gamma_{j+f+\sigma+1}(G) > \gamma_{2j+2f}(G)$ because of $j + f \geq \sigma + 1$. Clearly, $\langle x\rangle^G \leq \gamma_{j+f+lk}(G)$. It remains to show that $\langle x\rangle^G \geq \gamma_{j+f+lk+\sigma}(G)$. Write $y_1, \ldots y_\alpha$ for the representatives of $\gamma_{j+f+\sigma}(G)/\gamma_{j+f+\sigma+1}(G)$. They are of the form $\prod_{m=1}^h(\tilde{x}^{a_m})^{g_m}$, $a_m \in \mathbb{N}$, $g_m \in G$, some $h \in \mathbb{N}$. Applying the p^l-power map to $y_n\gamma_{j+f+\sigma+1}(G)$ yields $(\prod_{m=1}^h(\tilde{x}^{a_m})^{g_m})^{p^l}\gamma_{j+f+kl+\sigma+1}(G) = \prod(\tilde{x}^{a_m p^l})^{g_m}\gamma_{j+f+kl+\sigma+1}(G)$. These are generators of $\gamma_{j+f+lk+\sigma}(G)/\gamma_{j+f+lk+\sigma+1}(G)$. The representatives of this factor group lie in $\langle x\rangle^G$. q.e.d.

(X.2) Lemma. *Define G, j, k, σ as above. Define $z = max\{j, \sigma + k\}$, if $p \neq 2$ or $z = max\{j, \sigma + 2k\}$, if $p = 2$. Then all N in G satisfying $N \leq \gamma_z(G)$ are uniform.*

Proof. Let $N \lhd G$, $N \leq \gamma_z(G)$, $\tilde{s} := s(N) \leq \sigma$, choose n maximal such that $\gamma_{n+\tilde{s}}(G) \leq N \leq \gamma_n(G)$ for $n \geq z$. Then $[N, N] \leq [\gamma_n(G), \gamma_n(G)] \leq \gamma_{2n}(G)$ and $N^p \geq \gamma_{n+\tilde{s}}(G)^p = \gamma_{n+\tilde{s}+k}(G)$ if $p \neq 2$ or $N^4 \geq \gamma_{n+\tilde{s}}(G)^4 = \gamma_{n+\tilde{s}+2k}(G)$ if $p = 2$. Since $\gamma_{2n}(G) \leq \gamma_{n+\tilde{s}+k}(G)$ or $\gamma_{2n}(G) \leq \gamma_{n+\tilde{s}+2k}(G)$ the normal subgroup N is powerful. Since $\gamma_n(G)$ is uniform, it is also torsions free. Therefore all $N \lhd G$, $N \leq \gamma_z(G)$ are not only powerful but also torsion free and therefore uniform. q.e.d.

(X.3) Lemma. *With the above notation the set of normal subgroups $\{N | N \leq \gamma_{n+k}(G)$ and $N \not\geq \gamma_{n+2k}(G)\}$ equals the set $\{M^p | M \leq \gamma_n(G)$ and $M \not\geq \gamma_{n+k}(G)\}$ if $n \geq z$. In particular define $m := min\{n \in \mathbb{N} | N \not\leq \gamma_z(G), N \geq \gamma_n(G)\}$. Define $\alpha = max\{m - 1, z\}$. Then it follows that $|\gamma_{n+k+1}(G) : \mu_{n+k}(G)| = |\gamma_{n+1}(G) : \mu_n(G)|$ if $n \geq \alpha$.*

Proof. Let $n \geq j$. The p^l-te power of $N \leq \gamma_n(G)$ is a normal subgroup of $\gamma_{n+lk}(G)$. For every normal subgroup in $\gamma_{n+lk}(G)$ one gets by taking p^l -th roots a normal subgroup in $\gamma_n(G)$.
For the $n-th$ obliquity one has to consider only minimal normal subgroups, which are not contained in $\gamma_{n+1}(G)$. Therefore one only has to determine $\mu_l(G), \ldots, \mu_{l+k-1}(G)$, then $\mu_n(G)$ for $n \geq l + k - 1$ are given by taking p-th powers, if all minimal normal subgroups which appear in the intersection for $\mu_l(G), \ldots \mu_{l+k-1}(G)$ are uniform. Therefore for $n \geq max\{z, m-1\}$ one has a bijection between the normal subgroups in $\gamma_n(G)$ and $\gamma_{n+k}(G)$. Every normal subgroup which is not contained in $\gamma_z(G)$ contains

$\gamma_{\alpha+1}(G)$ and hence need not appear in the intersection for $\mu_n(G)$ with $n \geq \alpha - 1$. q.e.d.

If one carries out the calculation of the obliquity for $Q = G/\gamma_{c+1}(G)$ which is a finite group and if $\mu_{\alpha+k-1}(Q)$ contains a full set of representatives of $\gamma_c(G)/\gamma_{c+1}(G)$ then $\mu_\alpha(Q), \ldots, \mu_{\alpha+k-1}(Q)$ determine the period of the sequence $(\mu_i(G))$.

XI The theory behind the tables

In this chapter we describe the technical details of how the input for the program described earlier is obtained in order to compute the tables. There are essentially two issues: enumerate all relevant $\mathbb{Q}_p$-Lie algebras up to dimension 14 for $p \in \{2,3\}$ and secondly obtain generators for the Sylow pro-p-subgroups of their automorphism groups. The Lie algebras which are relevant are the simple ones and the semisimple ones having p^α isomorphic simple components for some $\alpha \in \mathbb{N}$. The simple Lie algebras $\mathcal{L}$ are absolutely simple over their centroid $C = \mathrm{End}_{\mathcal{L}}(\mathcal{L})$. The possible centroids are extensions of $\mathbb{Q}_2$ of degree 2, 3 and 4 and extensions of $\mathbb{Q}_3$ of degree 2, 3 and 4 which are given in Chapter IX.

We now enumerate the possible types of Lie algebras.

a) The relevant $\mathbb{Q}_p$-Lie algebras up to dimension 14

It is well known that a simple $\mathbb{Q}_p$-Lie algebra is absolutely simple over its centroid C. The classification (cf. [Sat 71]) leaves the following possibilities:

Notation: $\mathcal{K}_r = \mathcal{K}_r(K)$ is a central simple K-division algebra of index r for a p-adic field K. Let $sl_n(\mathcal{K}_r(K)) = \{x \in \mathcal{K}_r(K) | \text{ reduced trace } (x) = 0\}$.

dimension 3: Type A_1: $sl_2(\mathbb{Q}_p)$, $sl_1(\mathcal{K}_2(\mathbb{Q}_p))$.

dimension 6: Type A_1: $sl_2(K)$, $sl_1(\mathcal{K}_2(K))$ with $|K : \mathbb{Q}_p| = 2$.
Type $(A_1)^2$: $sl_2(\mathbb{Q}_2)^2$, $sl_1(\mathcal{K}_2(\mathbb{Q}_2))^2$.

dimension 8: Type A_2: $sl_3(\mathbb{Q}_p), sl_1(\mathcal{K}_3(\mathbb{Q}_p))$, $su_3(K, \mathbb{Q}_p)$ where $|K : \mathbb{Q}_p| = 2$.

dimension 9: Type A_1: $sl_2(K)$, $sl_1(\mathcal{K}_2(K))$ with $|K : \mathbb{Q}_p| = 3$.
Type $(A_1)^3$: $sl_2(\mathbb{Q}_3)^3$, $sl_1(\mathcal{K}_2(\mathbb{Q}_3))^3$.

dimension 10: Type B_2: $so_5(\phi_s, \mathbb{Q}_p)$, $so_5(\phi_{ns}, \mathbb{Q}_p)$ where ϕ_s and ϕ_{ns} can be chosen as the quadratic forms of the quadratic spaces $\langle 1\rangle \perp H \perp H$, $N \perp H$ where H is a hyberbolic plane over $\mathbb{Q}_p$, N the trace-0-subspace of $\mathcal{K}_2(\mathbb{Q}_p)$ with the quadratic form induced by the (reduced-) trace bilinear form of $\mathcal{K}_2(\mathbb{Q}_p)$.

dimension 12: Type A_1: $sl_2(K)$, $sl_1(\mathcal{K}_2(K))$ with $|K : \mathbb{Q}_p| = 4$.
Type $(A_1)^2$: $sl_2(K)^2$, $sl_1(\mathcal{K}_2(K))^2$ with $|K : \mathbb{Q}_p| = 2$.
Type $(A_1)^4$: $sl_2(\mathbb{Q}_2)^4$, $sl_1(\mathcal{K}_2(\mathbb{Q}_2))^4$.

dimension 14: Type G_2: $g_2(\mathbb{Q}_p)$.

b) Generators for the maximal p-adically simple groups

Define $\mathcal{O}_i^* := \{x \in \mathcal{O}^* | \nu_\pi(x-1) = i\}$ for $i \in \mathbb{N}$.

The following lemma is used at various places.

(XI.1) Lemma. *Let $r_{ij} \in \mathcal{O}_i^*$ such that $\langle r_{ij} \cdot \mathcal{O}_{i+1}^* | j = 1, \ldots, f \rangle = \mathcal{O}_i^*/\mathcal{O}_{i+1}^*$. Then $\mathcal{O}_1^*$ is (topologically) generated by $\{r_{ij} | j = 1, \ldots, f, i = 1, \ldots, e + \lfloor \frac{e}{p-1} \rfloor\}$.*

Proof. $\mathcal{O}_i^* \to \mathcal{O}_{i+e}^* : 1 + x \mapsto (1+x)^p$ is a homomorphism of $\mathcal{O}_{i+1}^*$ into $\mathcal{O}_{i+e+1}^*$. Expanding $(1+x)^p$ one gets terms of value 0, $e+i$, $e+2i$, $\ldots$, $e+(p-1)i$, ip. Hence for $p \cdot i > e+i$, i.e. $i > \frac{e}{p-1}$, one obtains an epimorphism of $\mathcal{O}_i^*/\mathcal{O}_{i+1}^*$ onto $\mathcal{O}_{i+e}^*/\mathcal{O}_{i+e+1}^*$. The claim follows. q.e.d.

Note, the proof also shows that usually a smaller generating set can be found.

Basically we have the following structure of $\mathrm{Aut}_{\mathbb{Q}_p}(\mathcal{L})$:

(XI.2) Lemma. *Let $\mathcal{L}$ be a simple $\mathbb{Q}_p$-Lie algebra with automorphism group $\mathrm{Aut}_{\mathbb{Q}_p}(\mathcal{L})$ and centroid $C := \mathrm{End}_{\mathcal{L}}(\mathcal{L})$.*

(i) $\mathrm{Aut}_{\mathbb{Q}_p}(\mathcal{L})$ is an extension of the group of C-Lie algebra automorphisms of $\mathcal{L}$ $\mathrm{Aut}_C(\mathcal{L})$ by a subgroup of $\mathrm{Aut}(C, \mathbb{Q}_p)$, the group of field automorphisms of C fixing $\mathbb{Q}_p$ pointwise.
If $\mathcal{L} = \mathcal{L}_1 \otimes_{\mathbb{Q}_p} C$ where $\mathcal{L}_1$ is a $\mathbb{Q}_p$-Lie algebra then

$$\mathrm{Aut}_{\mathbb{Q}_p}(\mathcal{L}) = \mathrm{Aut}_C(\mathcal{L}) \rtimes \mathrm{Aut}(C, \mathbb{Q}_p).$$

(ii) Let K be a (minimal) splitting field of $\mathcal{L}$, i.e. $\bar{\mathcal{L}} := K \otimes_C \mathcal{L}$ is a split K-Lie algebra of type $\Phi \in \{A_i, B_i, \ldots\}$ with Cartan subalgebra H. The following holds:

(a) Every $\alpha \in \mathrm{Aut}_C(\mathcal{L})$ extends uniquely to $\overline{\alpha} \in Aut_K(\overline{\mathcal{L}})$ i.e. $\alpha \mapsto \overline{\alpha}$ is a monomorphism,

(b) $\mathrm{Aut}_K(\bar{\mathcal{L}}) = G(\Phi, K) \cdot \tilde{\mathcal{R}} \cdot \mathcal{D}$ where $G(\Phi, K)$ is the Chevalley group with respect to the adjoint representation (cf. Chapter V) $\tilde{\mathcal{R}}$ consists of representatives of $\mathrm{Hom}(\mathbb{Z}\Phi, K^*)/\mathrm{Hom}(Q, K^*)$ and $\mathcal{D}$ is the finite group of diagram automorphisms acting faithfully on the Dynkin diagram of Φ.

Proof. (i) and the first part of (ii) is clear, the rest follows from [Ste 61]. q.e.d.

Therefore there are various reasons why it is not always straightforward to find generators for maximal $\tilde{p}$-groups. Here are names for the difficulties arising.

(i) the additional elements of $\mathcal{D}$, called diagram automorphisms,

(ii) representatives of $\mathrm{Hom}(\mathbb{Z}\Phi, K^*)/\mathrm{Hom}(Q, K^*)$ which are not contained in $G(\Phi, K)$ if $|Q : \mathbb{Z}\Phi| > 1$ (Q denotes the weight lattice), called rational automorphisms (or cf. [Car 72] diagonal automorphisms),

(iii) suitable settings, e.g. if one has the choice of different representations, we prefer those of low dimension even if one has to factor out by the centre.

c) $sl_n(K)$

Let Φ be a root system of type A_{n-1} and $\mathcal{L} = sl_n(K)$ with centroid $C = K$ a finite field extension of $\mathbb{Q}_p$. The group $GL_n(K)$ acts on $\mathcal{L}$ by conjugation. Since the kernel of this action is $K^* I_n$ one has $PGL_n(K)$ acting on $\mathcal{L}$. Denote by $\bar{\ }: \mathrm{GL}_n(K) \to \mathrm{PGL}_n(K)$ the natural epimorphism.
Use the notations for matrices $d_i(a_1, \ldots, a_n)$ as defined in (VI.10).

(XI.3) Lemma. *The K-automorphism group $\mathrm{Aut}_K(sl_n(K))$ is isomorphic to $PGL_n(K) \rtimes \mathcal{D}$ where $\mathcal{D}$ is the group of diagram automorphisms which is of order 2.*

Proof. The claim follows by (XI.2) (ii) and by the fact that $PGL_n(K)$ is a Chevalley group with the rational automorphisms on top. q.e.d.

A set of generators of the maximal $\tilde{p}$-group acting on $sl_n(K)$ (= Sylow pro-p-subgroup of $\mathrm{Aut}_{\mathbb{Q}_p}(sl_n(K))$) can be obtained as follows.

(XI.4) Lemma. *Let $E_{ij}(a) \in GL_n(K)$ be the matrix with 1 on the diagonal, a in position (i,j) and 0 everywhere else. Define $\mathcal{W}$ to be a $\mathbb{Z}_p$-basis of $\mathcal{O}_K$ and $\mathcal{B}$ a generating set of $\mathcal{O}_1^*$ (cf. XI.1). Define $\mathcal{E} := \{\overline{E_{i,i+1}(a)}, \overline{E_{n,1}(\pi a)} | i = 1, \ldots, n-1 \text{ and } a \in \mathcal{W}\}$. Define $r_n(\pi) := \overline{d_1(1, \ldots, 1, \pi)}$. Denote by $\mathcal{R}$ the set of rational automorphisms $\{r_n(\pi)^l, d_0(b, 1, \ldots, 1) | b \in \mathcal{B}\}$ if $n = p^k l$ and $p \nmid l$ for some $k \in \mathbb{N}$. Define* $F := \begin{pmatrix} & & 1 \\ & \cdot^{\cdot^{\cdot}} & \\ 1 & & \end{pmatrix}$. *The action of the diagram automorphism on the group generated by $\mathcal{E}$ is given by $\tilde{\ }: X \mapsto F X^{-tr} F$.*

a) A Sylow pro-p-subgroup of $\mathrm{Aut}_K(\mathcal{L}) \cong PGL_n(K) \rtimes \mathcal{D}$ is generated by

(i) $\mathcal{E}$ if $p \nmid n$ and p odd,

(ii) $\mathcal{E} \cup \{\tilde{\ }\}$ if $p \nmid n$ and p even,

(iii) $\mathcal{E} \cup \mathcal{R}$ if $n = p^k l$ for some $k, l \in \mathbb{N}$, $p \nmid l$ and p odd.

(iv) $\mathcal{E} \cup \mathcal{R} \cup \{\tilde{\ }\}$ if $n = p^k l$ for some $k, l \in \mathbb{N}$, $p \nmid l$ and p even.

b) Let Γ be the Sylow pro-p-subgroup of $\mathrm{Aut}(K, \mathbb{Q}_p)$ such that every $\sigma \in \Gamma$ satisfies $\pi\sigma \equiv \pi (\mathrm{mod}\ \pi^2 \mathcal{O})$. The Sylow pro-p-subgroup P of $\mathrm{Aut}_K(\mathcal{L})$ constructed in a) can be extended in a split way by Γ such that the elements of Γ act on the entries of the matrices.

Proof. a) Clearly, $\mathcal{E}$ generates a Sylow pro-p-subgroup of $\overline{SL_n(K)}$. The factor group $GL_n(K)/SL_n(K) \cong K^*$ and $A := GL_n(K)/(SL_n(K) \cdot K^* I_n) \cong K^*/(K^*)^n$. Hence if $p \nmid n$ one has that $\langle \mathcal{E} \rangle$ is a Sylow pro-p-subgroup of $PGL_n(K)$.
Otherwise, observe that the elements of $\mathcal{R}$ act on $\langle \mathcal{E} \rangle$. The natural image in A of the elements in $\mathcal{R}$ generate, via the above isomorphism, a Sylow pro-p-subgroup of

$K^*/(K^*)^n$. Therefore the set $\{\mathcal{E},\mathcal{R}\}$ generates a Sylow pro-p-subgroup of $PGL_n(K)$. If $p=2$ additionally the diagram automorphism acts on the Lie algebra by $D:\mathcal{L}\to\mathcal{L}: l\mapsto -Fl^{tr}F$ and induces $\tilde{}$ on the group generated by $\mathcal{E}$ or $\{\mathcal{E},\mathcal{R}\}$ preserving the chosen Sylow pro-p-subgroup.
b) W.l.o.g. choose Γ in the described way. The condition $\pi\sigma\equiv\pi \pmod{\pi^2\mathcal{O}}$ guarantees that P is mapped to itself by Γ. q.e.d.

d) $su_3(K,\mathbb{Q}_p)$

(XI.5) Lemma. *Let $K/\mathbb{Q}_p$ be a quadratic extension with $\mathbb{Q}_p$-automorphism $\bar{}$ of order 2 and* $F=\begin{pmatrix}0 & & 1\\ & \cdot^{\cdot^{\cdot}} & \\ 1 & & 0\end{pmatrix}\in K^{n\times n}$. *The special unitary Lie algebra* $\mathcal{L}= su_n(K,\mathbb{Q}_p):=\{X\in K^{n\times n}\,|\,XF+F\overline{X}^{tr}=0, trace(X)=0\}$ *has* $P\Gamma U_n(K,\mathbb{Q}_p)\rtimes\langle\bar{}\rangle$ *as its $\mathbb{Q}_p$-automorphism group, where* $\Gamma U_n(K,\mathbb{Q}_p):=\{g\in GL(n,K)\,|\,gF\bar{g}^{tr}=\lambda_g F$ *for some* $\lambda_g\in\mathbb{Q}_p^*\}$. *The factor group* $P\Gamma U_n(K,\mathbb{Q}_p)=\Gamma U_n(K,\mathbb{Q}_p)K^*I_n/K^*I_n$ *acts by conjugation on $\mathcal{L}$, i.e.* $gK^*I_n: x\in\mathcal{L}\mapsto gxg^{-1}$ *and* $\bar{}$ *acts by entry-wise application.*

Proof. Clearly $\tilde{\mathcal{L}}:=K\otimes_{\mathbb{Q}_p}\mathcal{L}\cong sl_n(K)$. By identifying $\mathcal{L}$ with $\mathbb{Q}_p\otimes_{\mathbb{Q}_p}\mathcal{L}\subset\tilde{\mathcal{L}}$, a $\mathbb{Q}_p$-basis of $\mathcal{L}$ yields a K-basis of the Lie algebra of trace zero matrices $sl_n(K)$. Let $D:\tilde{\mathcal{L}}\to\tilde{\mathcal{L}}: X\mapsto -FX^{tr}F$ denote the diagram automorphism on the Lie algebra. By (XI.2) each K-automorphism of $\mathcal{L}$ (viewed as a $\mathbb{Q}_p$-subalgebra of $\tilde{\mathcal{L}}$) lies in $PGL_n(K)\rtimes\langle D\rangle$. The natural $K-\mathcal{L}-$ (or $\tilde{\mathcal{L}}-$) module $K^{1\times n}$ is absolutely simple. Therefore by Schur's Lemma $\{F_1\in K^{n\times n}\,|\,XF_1=F_1(-\overline{X}^{tr})$ for all $X\in\mathcal{L}\}=KF$. Hence any $g\in GL_n(K)$ conjugating $\mathcal{L}$ into itself, must map F onto a multiple, i.e. $gF\bar{g}^{tr}=\lambda_g F$ or $g\in\Gamma U_n(K,\mathbb{Q}_p)$. Conversely every $g\in\Gamma U_n(K,\mathbb{Q}_p)$ induces an automorphism on $\mathcal{L}$. The diagram automorphism D maps $\mathcal{L}$ into itself and it restricts to the Galois automorphism $\bar{}$ on $\mathcal{L}$ acting entry-wise on the matrices. q.e.d.

(XI.6) Lemma.

(i) If in the last lemma n is odd, then $P\Gamma U_n(K,\mathbb{Q}_p)$ is isomorphic to $PU_n(K,\mathbb{Q}_p)$, i.e. $\mathrm{Aut}_{\mathbb{Q}_p}(su_n(K,\mathbb{Q}_p))\cong PU_n(K,\mathbb{Q}_p)\rtimes\langle\bar{}\rangle$.

(ii) Moreover, for arbitrary n, $PU_n(K,\mathbb{Q}_p)/PSU_n(K,\mathbb{Q}_p)\cong\mathrm{Ker}(N)/(\mathrm{Ker}(N))^n$ *where* $N:K^*\to\mathbb{Q}_p^*: x\mapsto x\bar{x}$ *is the norm map.*

Proof. (i) Let $\Gamma=\Gamma U_n(K,\mathbb{Q}_p)$ and $\lambda:\Gamma\to\mathbb{Q}_p^*: g\mapsto\lambda_g$ where $gF\bar{g}^{tr}=\lambda_g F$. Then $N(K^*)\le\Gamma\lambda\le\mathbb{Q}_p^*$, because λ maps the scalar matrices in Γ onto $N(K^*)$. Taking determinants one gets $\lambda_g^n=N(\det(g))$, i.e. $(\Gamma\lambda)^n\le N(K^*)$. But n is odd and by local class field theory (cf. [Neu 86] p. 42) $\mathbb{Q}_p^*/N(K^*)\cong C_2$ ($\cong\mathrm{Gal}(K/\mathbb{Q}_p)$), hence $\Gamma\lambda=N(K^*)$. Since $\mathrm{Ker}\,\lambda=U_n(K,\mathbb{Q}_p)$ one obtains $\Gamma U_n(K,\mathbb{Q}_p)=U_n(K,\mathbb{Q}_p)\cdot K^*I_n$, thus proving the first claim.

(ii) The map $\det: U_n(K,\mathbb{Q}_p)\to\mathrm{Ker}(N)$ is surjective, cf. [Tay 92] p. 115, and has $SU_n(K,\mathbb{Q}_p)$ as its kernel. Since the determinant of $\alpha I_n\in U_n(K,\mathbb{Q}_p)$ is α^n the last claim follows. q.e.d.

We proceed to determine generators for the Sylow pro-2-subgroup of $SU_3(K,\mathbb{Q}_2)$. We want to describe the group $SU_3(K,\mathbb{Q}_2)$ in terms of algebras with involutions, cf. Chapter VI. To fix some notation let K be a quadratic extension of $\mathbb{Q}_2$ with maximal order $o := \mathbb{Z}_2[\omega]$. Let $A = K^{3\times 3}$ and $\langle\sigma\rangle = \mathrm{Gal}(K,\mathbb{Q}_2)$. Denote the involution $\begin{pmatrix} a & b & c \\ d & e & f \\ g & h & i \end{pmatrix} \mapsto \begin{pmatrix} i^\sigma & f^\sigma & c^\sigma \\ h^\sigma & e^\sigma & b^\sigma \\ g^\sigma & d^\sigma & a^\sigma \end{pmatrix}$ on A by $^\circ$. The unitary group $U_3(K,\mathbb{Q}_2)$ is given in the following by $\{g \in A | g^\circ = g^{-1}\}$. A $^\circ$-invariant minimal hereditary order is given by $\Gamma := \begin{pmatrix} o & o & o \\ \pi & o & o \\ \pi & \pi & o \end{pmatrix}$. Its radical equals $J = J(\Gamma) = \begin{pmatrix} \pi & o & o \\ \pi & \pi & o \\ \pi & \pi & \pi \end{pmatrix}$.
Let $(J^i/J^{i+1})^- = \{x \in J^i/J^{i+1} \mid x^\circ = -x\}$.

(XI.7) Lemma. *Define $G := U_3(K,\mathbb{Q}_2) \cap (1+J)$ and $G_i := G \cap (1+J^i)$ for $i \in \mathbb{N}$. Then G has a filtration $G = G_1 > G_2 > \cdots G_n > \cdots$ with monomorphisms $(G_i/G_{i+1},\cdot) \to ((J^i/J^{i+1})^-,+)$.*

Proof. The monomorphisms are induced by the isomorphisms $(1+J^i)/(1+J^{i+1}) \cong J^i/J^{i+1}$ via $\overline{1+x} \mapsto \bar{x}$. For any $g = 1+x \in G_i$ one has

$$g^{-1} \equiv 1 - x \bmod (1+J^{i+1}).$$

Since $(1+x)^\circ = 1 + x^\circ$ it follows that $x^\circ \equiv -x \pmod{J^{i+1}}$ for $1+x \in G_i$. q.e.d.

The notation for matrices $d_i(a_0,a_1,a_2)$ given in (VI.10) will be used. The matrices are to be read as coset representatives in J^j/J^{j+1}.

First consider the case of a ramified splitting field K of $SU_3(K,\mathbb{Q}_2)$.
Let $\omega = \pi$ where π is a uniformising element of K. One wants to find generators of G_i/G_{i+1} that are mapped into the 2-dimensional $\mathbb{F}_2$-module $(J^i/J^{i+1})^-$. One observes that $(J^i/J^{i+1})^+ = (J^i/J^{i+1})^-$ because the characteristic of the residue class field is 2.
$(J^{3i+1}/J^{3i+2})^- = \langle d_1(\pi^i, -(\pi^\sigma)^i, 0), d_1(0,0,\pi^{i+1})\rangle$,
$(J^{3i+2}/J^{3i+3})^- = \langle d_2(\pi^i,0,0), d_2(0,\pi^{i+1},-(\pi^\sigma)^{i+1})\rangle$,
$(J^{3(i+1)}/J^{3(i+1)+1})^- = \langle d_0(\pi^{i+1},1,-(\pi^\sigma)^{i+1}), d_0(0,\pi^{i+1},0)\rangle$.
The map $G_i/G_{i+1} \to (J^i/J^{i+1})^- : 1+x \mapsto x$ is not surjective for every i. This is due to obstacles which can be seen by carrying out calculations of the following kind: Let $x \in (J^i/J^{i+1})^-$ and $y \in J^{i+1}$. Assume $g = 1+x+y \in G_i$. Then $1 = gg^{-1} = gg^\circ = (1+x+y)(1+x^\circ+y^\circ) = 1+x+x^\circ+xx^\circ+z_x(y)$, where $x+x^\circ \in J^{i+1}$ and $xx^\circ \in J^{2i}$ and $z_x(y)$ is a term in y and x. But it may happen that there is no solution $y \in J^{i+1}$ such that $x+x^\circ+xx^\circ+z_x(y) = 0$ as explained in detail for every case below. In such a case we have a contradiction to the above assumption $g \in G_i$.
The series of G_i/G_{i+1} for $i \in \mathbb{N}$ have the following dimensions over $\mathbb{F}_2$ ($\overline{}$ marking the repetition):

Case 1: $1,1,1,1,1,2,\overline{2,1,2,1,2,1}$ for the extensions with minimal polynomials x^2+2, x^2+6, x^2-2 or x^2-6.
Let $i \geq 1$. Consider G_{6i+2}/G_{6i+3}.

If $g = 1 + x + y$ with $x = d_2(2^i, 0, 0) \in (J^{6i+2}/J^{6i+3})^-$. Then $0 = gg^\circ - 1 \equiv x + x^\circ + y + y^\circ \equiv d_2(2^{i+1}, 0, 0) + y + y^\circ \pmod{J^{6(i+1)+3}}$ because all other summands lie in higher powers of J. Since $d_2(2^{i+1}, 0, 0) \in J^{6(i+1)+2}$ it is necessary for the existence of a solution to the equation that there is a $y \in J^{6i+3} - J^{6(i+1)+3}$ satisfying $y + y^\circ \equiv d_2(2^{i+1}, 0, 0) \pmod{J^{6(i+1)+3}}$. But no such solution exists.
Consider G_{6i+4}/G_{6i+5}. For $x = d_1(0, 0, 2^i)$ one finds that there is no solution for $g = 1 + x \pmod{J^{6i+5}}$ in an analogous way.
Consider $G_{6(i+1)}/G_{6(i+1)+1}$, i.e. $x = d_0(0, 2^{i+1}, 0) \in (J^{6(i+1)}/J^{6(i+1)+1})^-$. If $0 = gg^\circ - 1 \equiv x + x^\circ + y + y^\circ \pmod{J^{6(i+2)+1}}$ it follows that $x + x^\circ \in J^{6(i+2)}$. But since there exists no solution $y \in J^{6(i+1)+1} - J^{6(i+2)+1}$ to the congruence $x + x^\circ + y + y^\circ \equiv 0 \pmod{J^{6(i+3)}}$ it follows that $g = 1 + x \notin G_{6i+4}/G_{6i+5}$.
The arguments are very similar for $i = 0$ but one also has to consider the term xx° which causes a slightly different behaviour.
For all $g \equiv 1 + x \pmod{J^{i+1}}$ with $x \in (J^i/J^{i+1})^-$ and x not as above the congruences are solvable. This can be checked by calculations or follows from Lemma (XI.8) below.

Case 2: $1, 1, 2, 2, 1, 2, \overline{1, 2, 1, 2, 1, 2}$ for the extensions with minimal polynomials $x^2 + 2x + 2$, $x^2 + 2x + 6$.
Let $i \geq 1$. Consider G_{6i+1}/G_{6i+2}
Assume $g = 1 + x + y \in G_{6i+1}$ with $x = d_1(0, 0, 2^i\pi) \in (J^{6i+1}/J^{6i+2})^-$. Then $0 = gg^\circ - 1 \equiv x + x^\circ + y + y^\circ \equiv d_1(2^{i+1}, 0, 0) + y + y^\circ \pmod{J^{6i+5}}$. Hence $x + x^\circ \in J^{6i+4}$ there exists no solution to the equation.
For G_{6i+5}/G_{6i+6}, in particular $x = d_2(2^i\pi, 0, 0) \in (J^{6i+5}/J^{6i+6})^-$ the calculation is done in an analogue way and yields no solution for $g = 1 + x \pmod{J^{3i+6}}$.
Now consider G_{6i+3}/G_{6i+4}, i.e. $x = d_0(0, 2^i\pi, 0) \in J^{6i+3}$. If $0 = gg^\circ - 1 \equiv x + x^\circ + y + y^\circ \pmod{J^{6(i+1)+1}}$ hence $x + x^\circ = d_0(0, 2^{i+1}, 0) \in J^{6(i+1)}$. But the equation is not solvable.
The arguments are very similar for $i = 0$ but one also has to consider the term xx°.
For all $g \equiv 1 + x$ with $x \in (J^i/J^{i+1})^-$ and x not as above the congruences are solvable. This can be checked by calculations or follows from Lemma (XI.8).

Now consider the case of an unramified splitting field.
Let w be a primitive 3rd root of unity. The $\mathbb{F}_2$-modules $(J^i/J^{i+1})^-$ are 4-dimensional.
$(J^{3i+1}/J^{3i+2})^- = \langle d_1(2^i, -2^i, 0), d_1(2^i\omega, -2^i\omega^\sigma, 0), d_1(0, 0, 2^{i+1}), d_1(0, 0, 2^{i+1}\omega)\rangle$
$(J^{3i+2}/J^{3i+3})^- = \langle d_2(2^i, 0, 0), d_2(2^i\omega, 0, 0), d_2(0, 2^{i+1}, -2^{i+1}), d_2(0, 2^{i+1}\omega, -2^{i+1}\omega^\sigma)\rangle$
$(J^{3i+3}/J^{3i+4})^- = \langle d_3(2^{i+1}, 0, -2^{i+1}), d_3(2^{i+1}\omega, 0, -2^{i+1}\omega^\sigma), d_3(0, 2^{i+1}, 0), d_3(0, 2^{i+1}\omega, 0)\rangle$

But due to obstacles as in the ramified case one gets smaller dimensions for G_i/G_{i+1} :

Case 3: $\overline{3, 3, 3}$ for the unramified extension $(x^2 + x + 1)$.
For G_{3i+j}/G_{3i+j+1}, $1 \leq j \leq 3$ and $i \geq 1$ consider $g = 1 + x + y$ with $x \in (J^{3i+j}/J^{3i+j+1})^-$ then $0 = gg^\circ - 1 \equiv x + x^\circ + y + y^\circ \pmod{J^{3i+j+4}}$ If x fixed under $^\circ$ it follows that $x + x^\circ \in J^{3i+j+3}$. In particular these elements are $x = d_1(0, 0, 2^i)$, $x = d_2(2^i, 0, 0)$, $x = d_3(0, 2^i, 0)$. But there exists no y in $J^{3i+j+1} - J^{3i+j+4}$ satisfying $x + x^\circ \pmod{J^{3i+j+4}} \equiv y + y^\circ \pmod{J^{3i+j+4}}$. The

case $i = 0$ works out similarily. For all $g \equiv 1+x \pmod{J^{i+1}}$ with $x \in (J^i/J^{i+1})^-$ and x not as above the congruences are solvable. This can be checked by calculations or follows from Lemma (XI.8).

(XI.8) Lemma. *Using the notation for matrices $d_i(a_0, a_1, a_2)$ explained in (VI.10) the Sylow pro-2-subgroup G of $U_3(K, \mathbb{Q}_2)$ is generated by*

Case 1: extensions x^2+2, x^2+6, x^2-2, x^2-6
$I_3 + d_1(0,0,\pi), d_0(1+\pi, 1, (1+\pi^\sigma)^{-1}), I_3 + d_1(\pi,\pi,0) + d_2(-N(\pi)\cdot 2^{-1}, 0, 0)$, $d_0(-1,1,-1)$, $d_0(1,-1,1)$, $I_3 + d_1(0,0,-2) + d_2(0,2,-2)$, $d_0(1,c,1)$ *for* $cc^\sigma = 1$ *and* $c - 1 \equiv 2\pi \pmod 4$,

Case 2: extensions x^2+2x+2, x^2+2x+6
$I_3 + d_1(0,0,c) + d_2(0,\pi,-\pi^\sigma)$ *with* $c + c^\sigma = -N(\pi)$ *and* $\nu(c) = 1$,
$I_3 + d_2(c,0,0)$ *with* $trace(c) = 0, \nu(c) = 0$
$d_0(1+\pi, 1, (1+\pi^\sigma)^{-1})$, $d_0(1,c,1)$ *for* $c-1 = \pi \pmod 2$ *and* $N(c) = 1$, $d_0(-1,1,-1)$, $d_0(1,-1,1)$,

Case 3: unramified extension x^2+x+1
$I_3 + d_1(1,-1,0) + d_2(\omega,0,0), I_3 + d_1(0,0,-2-4\omega), d_0(1+2\omega, 1, (1+2\omega^\sigma)^{-1})$,
$I_3 + d_1(\omega, -\omega^\sigma, 0) + d_2(\omega,0,0)$, $d_0(3+2\omega, 1, (3+2\omega^\sigma))$.

The Sylow pro-2-subgroup of $\mathrm{Aut}_{\mathbb{Q}_2}(\mathcal{L})$ *is isomorphic to* $G/Z(G) \rtimes \langle\tilde{\sigma}\rangle$ *where* $Z(G)$ *denotes the centre of* G *and* $\tilde{\sigma}$ *is induced by the Galois automorphism* σ *acting entrywise on the matrices.*

Proof. Let H be topologically generated by the above elements. In all three cases the calculation of the lattice of H-invariant sublattices of $o^{1\times 3}$ (by a p-adic version of the sublattice algorithm cf. [PlP 77]) shows that there is a chain of distinguished sublattices through $o^{1\times 3}$. The Sylow pro-2-subgroup of the normaliser of the group acts on this chain. Define $H_i := H \cap G_i$. If the splitting field is ramified ($e = 2$) let $i_0 = 12$, otherwise ($e = 1$) $i_0 = 6$. By computation one checks that $H_i/H_{i+1} \cong G_i/G_{i+1}$ for $i \le i_0$. For $i > i_0$ one uses the map $x \mapsto x^2$ to prove the isomorphism by induction since for $g = 1+x+y \in H_i$ with $x \in J^i - J^{i+1}$ follows $g^2 = (1+x)^2 \equiv 1 + 2x \pmod{J^{3e+i+1}}$ and $2x \in J^{3e+i}$. Therefore all elements of the normaliser which fix this chain and lie in $U_3(K, \mathbb{Q}_p)$ have already been found. The rest follows from (XI.5) and (XI.6). q.e.d.

(XI.9) Remark. See also [Tit 79]. Apart from the residue class field being of characteristic 2 part of the difficulties here might result from the fact that the root system of $SU_3(K, \mathbb{Q}_p)$ is not reduced in the sense that some roots allow multiples which are also roots.

Now we determine the generators for $SU_3(K, \mathbb{Q}_3)$ for $p = 3$ and a ramified extension of $\mathbb{Q}_3$ as splitting field K. Since $(\mathrm{rad}\Lambda)^-$ is only semi-saturated (cf. VI.18) one has to choose a generator additionally to those which are images of $((\mathrm{rad}\Lambda)^1)^- - ((\mathrm{rad}\Lambda)^2)^-$ under the Cayley map $c_{LP} : x \mapsto \frac{1-x}{1+x}$ for constructing $SU_3(\mathbb{Q}_3[\pi]) \cap (1+J)$ ($\pi^2 = 3$ or $\pi^2 = -3$). From (XI.6) it follows that for generating PU_3 one also has to calculate extra diagonal elements which have a non-trivial entry in $\mathrm{Ker}N/(\mathrm{Ker}N)^3$. There is one element in this factor if $\pi^2 = 3$ or if the splitting field is unramified. There are two elements if $\pi^2 = -3$. Hence we have the following lemma.

(XI.10) Lemma. *The Sylow pro-3-subgroup of* $\mathrm{Aut}_{\mathbb{Q}_3}(\mathcal{L})$ *for a ramified splitting field is generated modulo central elements by the following matrices.*
If $\pi^2 = 3$: $(d_1(\pi,\pi,0))c_{LP}$, $(d_1(1,-1,0))c_{LP}$, $(d_1(0,0,\pi))c_{LP}$ *and* $d_0(1,a_1,1)$ *where* $a_1 = (1+3)(1+\pi)^\alpha$ *with* $\alpha = \frac{-2log_3(1+3)}{log_3(1-3)}$.
If $\pi^2 = -3$: $(d_1(\pi,\pi,0))c_{LP}$, $(d_1(1,-1,0))c_{LP}$, $(d_1(0,0,\pi))c_{LP}$, $d_0(1,a_1,1)$ *and* $d_0(1,a_2,1)$ *where* $a_1 = (1-3)^2(1+\pi)^\alpha$ *with* $\alpha = 2\frac{-2log_3(1-3)}{log_3(1+3)}$ *and* $a_2 = (1-3\pi)^2(1+\pi)^\beta$ *with* $\beta = 2\frac{-log_3(1+27)}{log_3(1+3)}$.

(XI.11) Lemma. *The Sylow pro-3-subgroup of* $\mathrm{Aut}_{\mathbb{Q}_3}(\mathcal{L})$ *with an unramified splitting field (e.g.* $K = \mathbb{Q}_3[\zeta_8]$*) is generated modulo central elements by the following matrices.*
$(d_1(1,-1,0))c_{LP}, (d_1(\omega,-\omega^\sigma,0))c_{LP}, (d_1(0,0,c))c_{LP}$ *with* $trace(c) = 0$ *and* $\nu(c) = 1$, $d_0(1,-1+\zeta_8-\zeta_8^2,1)$.

Proof. The generators follow straightforward from (VI.17) by adding $d_0(1,-1+\zeta_8-\zeta_8^2,1)$ as an extra generator to pass from SU_3 to PU_3 hence $z = -1+\zeta_8-\zeta_8^2$ generates $\mathrm{Ker}(N)/(\mathrm{Ker}(N))^3$. This last claim holds because the factor group $\mathrm{Ker}(N)$ modulo the torsion elements is isomorphic to $\mathbb{Z}_3$ since the index of $N(K^*)$ in $\mathbb{Q}_3$ is finite. Furthermore there is no 3-torsion in $\mathbb{Q}_3$ and no third root of z lies in $\mathbb{Z}_3[\zeta_8]$. q.e.d.

e) $sl_1(\mathcal{K}_2(K))$

Assume a division algebra D of dimension d^2 over its centre K is given in the following way, cf. [Rei 75]: Let w be a $(q^d-1)^{th}$ root of unity and $W = K[w]$. By $\pi_K \in K$ denote an element of valuation 1 in K. An automorphism θ generating $\mathrm{Gal}(W,K)$ maps $w \mapsto w^{q^r}$ for some r, $r \nmid n$. The elements of D are given as K-linear combinations of $(w^*)^i$ and $(\pi_D)^j$, $1 \le i,j \le n$, where $\alpha^* := d_0(\alpha,\alpha^\theta,\ldots,\alpha^{\theta^{n-1}})$ for any $\alpha \in W$ and $\pi_D := d_1(1,\ldots,1,\pi_K)$ which is an element of valuation 1 in D.
Consider the Lie algebra $\mathcal{L} = sl_1(D)$ consisting of those elements of D which have trace 0. Tensoring by W yields $W \otimes_K D = W^{d\times d}$ and $W \otimes_K \mathcal{L} \cong sl_d(W)$. Therefore we view D as K-subalgebra of $W^{d\times d}$.
The automorphism $\theta \in \mathrm{Gal}(W,K)$ applied entry-wise to the matrices induces an automorphism on D.
Now we want to exhibit how to extend the $\mathbb{Q}_p$-automorphisms of the centre K to an action on D.

(XI.12) Lemma. *Let* K *be an extension of* $\mathbb{Q}_p$ *with* $|K:\mathbb{Q}_p| \le 4$ *and* W *an unramified extension of* K *of degree* d. *Define* $F := Fix_{\mathrm{Aut}_{\mathbb{Q}_p}(K)}(K)$ *a fixed subfield in* K. *Then,* W *is a Galois extension over* F *with an abelian Galois group.*

Proof. Define $F_u < K$ to be the maximal unramified subfield over F of the extension K over F. Since K is a Galois extension over F with abelian Galois group

$\mathrm{Gal}(K,F)$ it follows that $W = K\tilde{F}$ where $\tilde{F}$ is the unramified extension of degree d of F_u. The Galois group of $\tilde{F}$ over F is cyclic and the Galois group of K over F is abelian of degree less or equal to 4. The claim follows since in the Galois group $G = \mathrm{Gal}(W,F)$ the subgroups corresponding to K and $\tilde{F}$ are normal with trivial intersection and have abelian factor groups in G. q.e.d.

(XI.13) Lemma. *With the above notation and* $|K : \mathbb{Q}_p| \leq 4$ *it follows that* $\mathrm{Aut}_{\mathbb{Q}_p}(D) = \mathrm{Aut}_F(D) \rhd \mathrm{Aut}_K(D)$ *and* $\mathrm{Aut}_{\mathbb{Q}_p}(D)/\mathrm{Aut}_K(D) \cong \mathrm{Aut}_{\mathbb{Q}_p}(K) \cong \mathrm{Gal}(K,F)$. *In particular, every* $\mathbb{Q}_p$*-automorphism of the centre* K *of* D *can be extended to a* $\mathbb{Q}_p$*-automorphism of* D.

Proof. Since the automorphism group of D acts on K it only remains to show how the extension of the automorphisms $\mathrm{Aut}(K,\mathbb{Q}_p)$ works out. Let $[i \cdot d^{-1}]$ be the Hasse invariant of D over K and $\theta \in \mathrm{Gal}(W,K)$ the i-th power of the Frobenius automorphism. Applying Lemma (XI.12) one gets that every $\psi \in \mathrm{Aut}(K,\mathbb{Q}_p) \cong \mathrm{Gal}(K,F)$ lifts to d elements $\psi_1,\ldots,\psi_d \in \mathrm{Gal}(W,F)$ with $\psi_i = \theta^{i-1}\psi_1$ for $i = 2,\ldots,d$. Applying ψ_i to D yields D^{ψ_i} which has the same Hasse invariant as D since ψ_1 and θ commute. Hence there exists a matrix in $W^{d\times d}$ which conjugates D^{ψ_i} to D. A matrix for this purpose is given by $M := d_0(1, n_\psi, \ldots, n_\psi^{1+\theta+\cdots+\theta^{d-2}})$ where $n_\psi \in W$ is a solution of $N_{W/K} n_\psi = \frac{\pi_K^\psi}{\pi_K}$. The conjugation operation by M is denoted by κ_M. Then, $\psi_1\kappa_M$ is a $\mathbb{Q}_p$-automorphism of D. q.e.d.

(XI.14) Lemma. *Let* K *be the centre of the division algebra* D *and let* $\mathcal{L}$ *be* $sl_1(D)$. *Then* $\mathrm{Aut}_K(\mathcal{L}) = PGL_1(D)$.

Proof. If $d > 2$ then any automorphism of $\mathcal{L}$ is induced by conjugation of an element of $PGL_1(D)$ since the two epimorphic images of the universal enveloping algebra $U(\mathcal{L})$ namely D and D^{op} which come from the two representations of $sl_d(W)$ of lowest degree are not isomorphic as algebras. If $d = 2$ the two division algebras of this degree and the two representations of the universal enveloping algebra $U(\mathcal{L})$ become isomorphic such that one also gets all automorphisms by conjugation of $PGL_1(D)$. The automorphism $\theta \in \mathrm{Gal}(W,K)$ applied entry-wise on $PGL_1(D)$ acts in the same way as $\pi_D \in PGL_1(D)$ by conjugation. q.e.d.

(XI.15) Lemma.

a) *Denote by* $\overline{} : GL_1(D) \to PGL_1(D)$ *the natural epimorphism. Let* π *be a uniformising element of* $\mathcal{O}_K$. *Let* e *be the ramification index of the centre* K *of* D *and* $\mathcal{W}$ *a* $\mathbb{Z}_p$*-basis of* $\mathcal{O}_W$. *Define* $U(s) = \{\overline{1 + \pi_D^j \alpha^*} | \alpha \in \mathcal{W},\ j = 1,\ldots,s\}$.

   The Sylow pro-p-subgroup of $PGL_1(D)$ *is generated by*

   (i) $U(de + \lfloor \frac{de}{p-1} \rfloor)$ *if* $p \not| d$,

   (ii) $U(de + \lfloor \frac{de}{p-1} \rfloor) \cup \{\overline{\pi_D^l}\}$ *if* $d = p^k l$ *and* $p \not| l$.

b) *Let* $|K : \mathbb{Q}_p| \leq 4$. *Define* Γ *to be the Sylow p-subgroup of* $\mathrm{Aut}_{\mathbb{Q}_p}(K)$ *such that for* $\psi \in \Gamma$ *holds* $\pi^\psi \equiv \pi (\mathrm{mod}\ \pi^2)$. *The Sylow pro-p-subgroup* $\mathrm{Aut}_{\mathbb{Q}_p}(\mathcal{L})$ *is isomorphic to* $PGL_1(D)$ *extended by the* Γ *described in Lemma (XI.13).*

Proof. a) Let Λ be the maximal order of D. Let ω be a primitive $(p^{fd}-1)^{th}$ root of unity where f is the inertia degree of K. It follows that $\Lambda = \sum_{i=0}^{d-1} R[\omega]\pi_D^i$ with valuation ring R of K ([Rei 75] p. 146). The Jacobson radical is given by $J(\Lambda) = \sum_{i=1}^{d} R[\omega]\pi_D^i$. The Sylow pro-$p$-subgroup G of $GL_1(D)$ is $1+J(\Lambda)$. There is a filtration of the group G by normal subgroups $G_i = 1+(J(\Lambda))^i$ such that there is an embedding $G_i/G_{i+1} \hookrightarrow J(\Lambda)^i/J(\Lambda)^{i+1}$ via $\overline{1+x} \mapsto \bar{x}$. Let H be generated by one of the sets U according to the specific case. Define H_i by $H_i := H \cap (1+J(\Lambda)^i)$. For $i \leq de + \lfloor \frac{de}{p-1} \rfloor$ H_i/H_{i+1} maps surjective to $J(\Lambda)^i/J(\Lambda)^{i+1}$. The map $y \mapsto y^p$ has the effect that for an element $1+x \in H_i$ with $x \in J^i - J^{i+1}$ the p-th power $(1+x)^p \equiv 1+px (\text{mod } J^{i+de+1})$ with $px \in J^{i+de}$ and $px (\text{mod } J^{i+de+1})$ is non-trivial. Therefore it follows by induction that for every $i > de + \lfloor \frac{de}{p-1} \rfloor$ every quotient $H_i/H_{i+1} \cong J(\Lambda)^i/J(\Lambda)^{i+1}$. This proves that H is a full Sylow pro-p-subgroup of $GL_1(D)$. Now consider $PGL_1(D)$. If d is divisible by p, the matrix $\overline{\pi_D}$ is the only missing representative for a generating set. For the Sylow pro-p-subgroup of $PGL_1(D)$ one has to take the l-th power of this matrix where $n = l \cdot p^u$ with l not divisible by p.
b) The choice of Γ makes sure that Γ normalises the chosen Sylow pro-p-subgroup of $PGL_1(D)$. The claim follows from Lemma (XI.13). q.e.d.

Remark. The given number of generators is far too big. In every specific case one easily reduces this number by also taking into account what one gets by commutators.

f) $so_5(\mathbb{Q}_p)$, split case

(XI.16) Lemma. *The $\mathbb{Q}_p$-Lie algebra $\mathcal{L} = so_5(\mathbb{Q}_p)$ is generated by*
$\bar{X}_\alpha = d_1(0,2,-1,0,0)$, $\bar{X}_\alpha = d_1(1,0,0,-1,0)$, *and* $\bar{X}_{-2\alpha-\beta} = d_2(0,0,0,1,-1)$ *as a Lie subalgebra of* $sl_5(\mathbb{Q}_p)$. *For* $p \neq 2$ *the generators for the maximal Sylow pro-p-subgroup of* $\mathrm{Aut}_{\mathbb{Q}_p}(\mathcal{L})$ *are*
$x_\alpha(1) = exp(\bar{X}_\alpha), x_\beta(1) = exp(\bar{X}_\beta)$ *and* $x_{-2\alpha-\beta}(p) = exp(p\bar{X}_{-2\alpha-\beta})$.

Proof. This is an immediate consequence of (V.3). q.e.d.

(XI.17) Remark. For $p = 2$ the situation is more complicated because in the first instance one has to deal with obstacle (i) (see after (V.3)) since one of the Cartan numbers is 2. To overcome this one has to add an extra generator for example $x_{2\alpha+\beta}(1) = exp(\bar{X}_{2\alpha+\beta})$. But also obstacle (ii) occurs that is $|Q : \mathbb{Z}\Phi| = 2$. One can observe that all generators given so far have spinor norm $spin(x) = 1$.
Denote by $\bar{G} := \{x \in SO_5(\mathbb{Q}_2) | spin(x) = 1\}$. One has $SO_5(\mathbb{Q}_2)/\bar{G} \cong \mathbb{Q}_2^*/(\mathbb{Q}_2^*)^2$ (see [OMe 73] p. 141). As described in [OMe 73] p. 242 one can transform the matrices for a Sylow pro-p- subgroup of $SO_5(\mathbb{Q}_2)$ to a basis such that one gets an integral representation.

A different approach is the following.
One takes the projective representation of $SO_5(\mathbb{Q}_p)$ on the 4-dimensional module called V utilising $B_2 = C_2$. The construction of a full Sylow pro-2-subgroup of the automorphism group of the Lie algebra is done in the next lemma. Together with some outer automorphisms the group representation on V can be transformed to the 5-dimensional irreducible representation by factoring out the the trivial module from

the exterior square tensor module $\wedge^2 V$.

Consider $SO_5(\mathbb{Q}_2)$ in the 4-dimensional projective representation in terms of algebras with involution (see Chapter VI). Let $A = \mathbb{Q}_2^{4\times 4}$ and $F = \begin{pmatrix} 0 & 0 & 0 & 1 \\ 0 & 0 & 1 & 0 \\ 0 & -1 & 0 & 0 \\ -1 & 0 & 0 & 0 \end{pmatrix}$.

Denote the involution

$$a = \begin{pmatrix} a_{11} & a_{12} & a_{13} & a_{14} \\ a_{21} & a_{22} & a_{23} & a_{24} \\ a_{31} & a_{32} & a_{33} & a_{34} \\ a_{41} & a_{42} & a_{43} & a_{44} \end{pmatrix} \mapsto Fa^{tr}F^{-1} = \begin{pmatrix} a_{44} & a_{34} & -a_{24} & -a_{14} \\ a_{43} & a_{33} & -a_{23} & -a_{13} \\ -a_{42} & -a_{32} & a_{22} & a_{12} \\ -a_{41} & -a_{31} & a_{21} & a_{11} \end{pmatrix} \text{ on } A \text{ by } {}^\circ.$$

An $^\circ$-invariant minimal hereditary order denoted by Γ is of the form $\begin{pmatrix} o & o & o & o \\ 2 & o & o & o \\ 2 & 2 & o & o \\ 2 & 2 & 2 & o \end{pmatrix}$.

Its radical is of the form $J = J(\Gamma) = \begin{pmatrix} 2 & o & o & o \\ 2 & 2 & o & o \\ 2 & 2 & 2 & o \\ 2 & 2 & 2 & 2 \end{pmatrix}$. One has $so_5(\mathbb{Q}_2) \cong sp_4(\mathbb{Q}_2)$, where $sp_4(\mathbb{Q}_2) = \{a \in A | a^\circ = -a\}$. Define the symplectic group

$$Sp_4(\mathbb{Q}_2) := \{x \in A | x^{-1} = x^\circ\}.$$

(XI.18) Lemma. *With the above notation on has*

a) $(J^1/J^2)^- = \langle d_1(1,0,-1,0), d_1(0,1,0,0), d_1(0,0,0,2)\rangle$,
$(J^2/J^3)^- = \langle d_2(1,1,0,0), d_2(0,0,2,2)\rangle$,
$(J^3/J^4)^- = \langle d_3(1,0,0,0), d_3(0,2,0,-2), d_3(0,0,2,0)\rangle$,
$(J^4/J^5)^- = \langle d_0(2,0,0,-2), d_0(0,2,-2,0)\rangle$,
$(J^{4k+1}/J^{4k+2})^- = 2^k(J^1/J^2)^-$, *dimension* 3,
$(J^{4k+2}/J^{4k+3})^- = 2^k(J^2/J^3)^-$, *dimension* 2,
$(J^{4k+3}/J^{4k+4})^- = 2^k(J^3/J^4)^-$, *dimension* 3,
$(J^{4k+4}/J^{4k+5})^- = 2^k(J^4/J^5)^-$, *dimension* 2.

b) $G := Sp_4(\mathbb{Q}_2) \cap (1+J)$ *has a filtration by normal subgroups* $G_i := G \cap (1+J^i)$ $G_1 = G > G_2 > \cdots$ *where* $(G_i/G_{i+1}, \cdot)$ *is embedded into* $((J^i/J^{i+1})^-, +)$ *via* $\overline{1+x} \mapsto \overline{x}$.

c) The group G is generated by $\mathcal{H} := \{I_4 + d_1(1,0,-1,0), I_4 + d_1(0,1,0,0), I_4 + d_1(0,0,0,2), I_4 + d_3(1,0,0,0), I_4 + d_3(0,0,2,0)\}$.
The group G and representatives of the rational automorphisms $\mathcal{R} = \{d_0(1,1,-1,-1), d_0(1,1,3,3), d_2(1,1,2,2)\}$ *act by conjugation on* $\mathcal{L}$. *The kernel of this action consists of scalar matrices. The Sylow pro-2-subgroup of* $\mathrm{Aut}_{\mathbb{Q}_2}(\mathcal{L})$ *is generated by* $\mathcal{H}$ *and* $\mathcal{R}$ *modulo scalar matrices.*

Proof. a) and b) can be checked by straightforward calculations analogous to Lemma (XI.7) and the calculation which follows Lemma (XI.7). The maps from $G_i/G_{i+1} \to (J^i/J^{i+1})^-$ are surjective in this case.
c) Let H be generated by $\mathcal{H}$. Define $H_i := H \cap G_i$. For $i \leq 8$ it is checked by

calculations that $H_i/H_{i+1} \cong (J^i/J^{i+1})^-$. For $i > 8$ these isomorphisms follow by induction applying the map $x \mapsto x^2$ to the elements hence for $g = 1 + x + y \in H_i$ with $x \in J^i - J^{i+1}$ follows $g^2 = (1 + x + y)^2 \equiv 1 + 2x \pmod{J^{i+5}}$ and $2x \in J^{i+4}$ is not trivial modulo J^{i+5}. All other terms lie in higher powers of J.
Any $\mathbb{Q}_2$-automorphism of $\mathcal{L}$ is induced by conjugation of an element $\phi \in PGL_4(\mathbb{Q}_2)$ acting on $\mathbb{Q}_2^{1\times 4}$ since there is only one faithful irreducible 4-dimensional representation for the Lie algebra B_2. The lattice of H-invariant lattices of $o^{1\times 4}$ consists of a chain. The Sylow pro-2-subgroup of the normaliser of G in $PGL_4(\mathbb{Q}_2)$ acts on this chain. Its elements either lie in $Sp_4(\mathbb{Q}_2) \cap (1 + J)$ or act non-trivially on the form F. Any automorphism ϕ of $\mathcal{L}$ acts on $\{A \in \mathbb{Q}_2^{4\times 4} | lA = -Al^{tr} \text{ for all } l \in \mathcal{L}\} = \mathbb{Q}_2 F$. The ones acting non-trivially on the form F map F to a multiple and hence are similitudes. The factor group $Sim(\mathbb{Q}_2, ^\circ)/(Sp_4(\mathbb{Q}_2) \cdot \mathbb{Q}_2 I_4)$ embeds into $\mathbb{Q}_2^*/(\mathbb{Q}_2^*)^4$ by mapping the elements to their scaling factor. The group of similitudes $Sim(\mathbb{Q}_2, ^\circ)$ contains $\bar{G} = G \cdot \mathbb{Q}_2^* I_4$. Let N be the the normaliser $N_{Sim(\mathbb{Q}_2,^\circ)}(\bar{G})$ and $U := Sp_4(\mathbb{Q}_2) \cdot \mathbb{Q}_2 I_4$. Then $Z := N/(N \cap Sp_4(\mathbb{Q}_2)\mathbb{Q}_2 I_4)$ embeds into $\mathbb{Q}_2^*/(\mathbb{Q}_2^*)^4$. Since $\bar{G}$ is of finite p-prime index in $N_U(\bar{G}) = N_{Sim(\mathbb{Q}_2,^\circ)}(\bar{G}) \cap U$ it follows that $\bar{G}$ can be extended by Z. Such an element has determinant y^{2k} for $k \in \mathbb{N}$ and $y \in \mathbb{Q}_p$ since $det(aFa^{tr}) = det(yF) = y^4 F$ and therefore $det(a)^2 = y^4$. Hence as representatives modulo scalar matrices one has $d_0(1, 1, -1, -1), d_0(1, 1, 3, 3)$ and $d_2(1, 1, 2, 2)$ as generators. q.e.d.

g) $so_5(\mathbb{Q}_p)$, non-split case

Let K be an unramified extension of $\mathbb{Q}_2$ of degree 2, w a 3rd root of unity $\mathrm{Gal}(K, \mathbb{Q}_2) = \langle \sigma \rangle$. Denote by $\mathcal{Q} := \mathcal{K}_2(\mathbb{Q}_2)$ the quaternions over $\mathbb{Q}_2$ given as

$$\begin{aligned} \mathcal{Q} &= \{\begin{pmatrix} a & b \\ 2b^\sigma & a^\sigma \end{pmatrix} | a, b \in \mathbb{Q}_2[\omega]\} \\ &= \langle I = \begin{pmatrix} 1 & 0 \\ 0 & 1 \end{pmatrix}, i = \begin{pmatrix} \omega & 0 \\ 0 & \omega^\sigma \end{pmatrix}, j = \begin{pmatrix} 0 & 1 \\ 2 & 0 \end{pmatrix}, k = ij \rangle. \end{aligned}$$

The map $\bar{\ } : \mathcal{Q} \to \mathcal{Q} : \begin{pmatrix} a & b \\ 2b^\sigma & a^\sigma \end{pmatrix} \mapsto \begin{pmatrix} a^\sigma & -b \\ -2b^\sigma & a \end{pmatrix}$ defines an involution on $\mathcal{Q}$.
The $\mathbb{Z}_2$-maximal order of $\mathcal{Q}$ is called Ω and its radical $J(\Omega)$.

On $\mathcal{Q}^{2\times 2}$ we have the involution $^\circ : \begin{pmatrix} A & B \\ C & D \end{pmatrix} \mapsto \begin{pmatrix} \bar{D} & \bar{B} \\ \bar{C} & \bar{A} \end{pmatrix}$ given by the $\bar{\ }$-hermitian form $\begin{pmatrix} 0 & 1 \\ 1 & 0 \end{pmatrix}$ and a minimal $^\circ$-invariant hereditary order $\Gamma := \begin{pmatrix} \Omega & \Omega \\ J(\Omega) & \Omega \end{pmatrix}$.
The radical $J(\Gamma)$ is given by $\begin{pmatrix} J(\Omega) & \Omega \\ J(\Omega) & J(\Omega) \end{pmatrix}$.
It follows that the factors $J(\Gamma)^i/J(\Gamma)^{i+1} \cong \mathbb{F}_4^2$.
Note, because the characteristic of the residue field is 2 it follows that $(J(\Gamma)^i/J(\Gamma)^{i+1})^- = (J(\Gamma)^i/J(\Gamma)^{i+1})^-$.

<u>(XI.19) Lemma.</u> *With the above notation and $\mathcal{L} = \{x \in \mathcal{Q}^{2\times 2} | x^\circ = -x\}$ which is of type B_2 the following holds.*

a) $(J(\Gamma)/J(\Gamma)^2)^- = \langle \begin{pmatrix} 0 & I \\ 0 & 0 \end{pmatrix} \begin{pmatrix} 0 & 0 \\ j & 0 \end{pmatrix} \begin{pmatrix} 0 & 0 \\ k & 0 \end{pmatrix} \rangle,$

$(J(\Gamma)^2/J(\Gamma)^3)^- = \langle \begin{pmatrix} j & 0 \\ 0 & -\bar{j} \end{pmatrix} \begin{pmatrix} k & 0 \\ 0 & -\bar{k} \end{pmatrix} \rangle,$

$(J(\Gamma)^3/J(\Gamma)^4)^- = \langle \begin{pmatrix} 0 & j \\ 0 & 0 \end{pmatrix} \begin{pmatrix} 0 & k \\ 0 & 0 \end{pmatrix} \begin{pmatrix} 0 & 0 \\ 2I & 0 \end{pmatrix} \rangle,$

$(J(\Gamma)^4/J(\Gamma)^5)^- = \langle \begin{pmatrix} 2I & 0 \\ 0 & -2I \end{pmatrix} \begin{pmatrix} 2i & 0 \\ 0 & -2\bar{i} \end{pmatrix} \rangle,$

$(J(\Gamma)^{4k+1}/J(\Gamma)^{4k+2})^- = 2^k(J(\Gamma)^1/J(\Gamma)^2)^-$, *dimension* 3,

$(J(\Gamma)^{4k+2}/J(\Gamma)^{4k+3})^- = 2^k(J(\Gamma)^2/J(\Gamma)^3)^-$, *dimension* 2,

$(J(\Gamma)^{4k+3}/J(\Gamma)^{4k+4})^- = 2^k(J(\Gamma)^3/J(\Gamma)^4)^-$, *dimension* 3,

$(J(\Gamma)^{4k+4}/J(\Gamma)^{4k+5})^- = 2^k(J(\Gamma)^4/J(\Gamma)^5)^-$, *dimension* 2.

b) The subgroup $G := U_2(\mathcal{Q}, F) \cap (1 + J(\Gamma))$ *has a filtration* $G_i := G \cap (1 + J(\Gamma)^i)$ *with* $G_1 = G > G_2 > \cdots > G_i > \cdots$ *where* G_i/G_{i+1} *embeds into* $(J(\Gamma)^i/J(\Gamma)^{i+1})^-$ *via* $1 + \bar{x} \mapsto \bar{x}$.

c) The set $\mathcal{H}_1 := \{ \begin{pmatrix} I & I \\ 0 & I \end{pmatrix}, \begin{pmatrix} I & 0 \\ j & I \end{pmatrix}, \begin{pmatrix} I & 0 \\ k & I \end{pmatrix}, \begin{pmatrix} I & j \\ 0 & I \end{pmatrix}, \begin{pmatrix} I & k \\ 0 & I \end{pmatrix} \}$ *and the rational automorphisms*
$\mathcal{H}_2 := \{ \begin{pmatrix} I & 0 \\ 0 & -I \end{pmatrix}, \begin{pmatrix} I & 0 \\ 0 & 3I \end{pmatrix}, \begin{pmatrix} j & 0 \\ 0 & j \end{pmatrix} \}$ *generate a maximal Sylow pro-2-subgroup of* $\mathrm{Aut}_{\mathbb{Q}_2}(\mathcal{L})$.

Proof. a) and b) can be checked by straightforward calculations analogous to Lemma (XI.7).
c) Any $\mathbb{Q}_2$-automorphism of $\mathcal{L}$ is induced by conjugation of an element $\phi \in GL_2(\mathcal{Q})$ since the associative span $\langle \mathcal{L} \rangle_{asso} = \mathcal{Q}^{2\times 2}$ is the only epimorphic image of the universal enveloping algebra $U(\mathcal{L})$ in the requested dimension as seen from representation theory of B_2. Note, by the theorem of Skolem and Noether it follows that the $\mathbb{Q}_2$-automorphisms of $\mathcal{Q}^{2\times 2}$ are inner and therefore embed into $PGL_2(\mathcal{Q})$. Any automorphism ϕ maps $\{A \in \mathcal{Q}^{2\times 2} | lA = -A\bar{l}^{tr} \text{ for all } l \in \mathcal{L}\} = \mathbb{Q}_2 \begin{pmatrix} 0 & 1 \\ 1 & 0 \end{pmatrix}$ to itself. Henceforth $\mathrm{Aut}_{\mathbb{Q}_2}(\mathcal{L})$ maps into the similitudes modulo their centre $PSim(\mathcal{Q}, ^\circ)$. The set $\mathcal{H}_1$ is a generating set for the Sylow pro-2-subgroup G of $U_2(\mathcal{Q}, F)$. This can be proved by considering $H := \langle \mathcal{H}_1 \rangle$, $H_i := H \cap G_i$. For $i \leq 8$ one checks that the factors H_i/H_{i+1} are isomorphic to $(J(\Gamma)^i/J(\Gamma)^{i+1})^-$. The same follows by induction for $i > 8$ using $(1 + x)^2 \equiv 1 + 2x \pmod{J(\Gamma)^{i+5}}$ and $2x \in J(\Gamma)^{i+4}$ is non-trivial modulo $J(\Gamma)^{i+5}$ if $x \in J(\Gamma)^i/J(\Gamma)^{i+1}$. One finds that G fixes a chain of Ω-lattices. Define $U := U_2(\mathcal{Q}) \cdot \mathbb{Q}_2 I_2$. The factor group $Sim(\mathcal{Q}, ^\circ)/U$ embeds into $\mathbb{Q}_2^*/(\mathbb{Q}_2^*)^2 \cong C_2^3$ by mapping the elements to their scaling factor. The group of similitudes $Sim(\mathcal{Q}, ^\circ)$ contains $\bar{G} = G \cdot \mathbb{Q}_2^* I_2$. Let N be the normaliser $N_{Sim(\mathcal{Q}, ^\circ)}(\bar{G})$. Then $Z := N/(N \cap U)$ embeds into $\mathbb{Q}_2^*/(\mathbb{Q}_2^*)^2$. Since $\bar{G}$ is of finite p-prime index in $N_U(\bar{G}) = N_{Sim(\mathcal{Q}, ^\circ)}(\bar{G}) \cap U$ it follows that $\bar{G}$ can be extended by Z. In particular, the elements of $\mathcal{H}_2$ are representatives of this factor where the first 2 elements act trivially on the chain of sublattices and the third one non-trivially. q.e.d.

h) $g_2(\mathbb{Q}_p)$

(XI.20) Lemma.

(i) *The* $\mathbb{Q}_p$*-Lie algebra* $\mathcal{L}$ *of type* G_2 *is generated by* $\bar{X}_\alpha = d_1(0,-1,0,0,1,0,0)$, $\bar{X}_\beta = d_1(1,0,-1,2,0,-1,0)$ *and* $\bar{X}_{-2\alpha-3\beta} = d_2(0,0,0,0,0,1,-1)$ *as a Lie subalgebra of* $sl_7(\mathbb{Q}_p)$.

(ii) *For* $p \neq 2,3$ *the generators for the maximal Sylow pro-p-subgroup of* $\mathrm{Aut}_{\mathbb{Q}_p}(\mathcal{L})$ *are given by* $x_\alpha(1) = exp(\bar{X}_\alpha), x_\beta(1) = exp(\bar{X}_\beta)$ *and* $x_{-2\alpha-3\beta}(p) = exp(p\bar{X}_{-2\alpha-3\beta})$. *For* $p = 2$ *one needs additionally*

$$x_{\alpha+2\beta}(2) = exp(2\bar{X}_{\alpha+2\beta}) \text{ where } \bar{X}_{\alpha+2\beta} = d_3(-1,1,-1,2,0,0,0).$$

For $p = 3$ *one needs additionally*

$$x_{\alpha+3\beta}(3) = exp(3\bar{X}_{\alpha+3\beta}) \text{ where } \bar{X}_{\alpha+3\beta} = d_4(1,0,-1,0,0,0,0).$$

Proof. This is an immediate consequence of (V.3) and dealing with obstacle (i), i.e. that certain Cartan numbers are divisible by 2 or 3. q.e.d.

XII Tables

The following tables give computer generated information about the lower central series and normal subgroups of the maximal insoluble p-adically simple groups for $p = 2$ and $p = 3$ up to dimension 14. For comparison it also includes theoretically generated information on the corresponding groups, in case they exist, for primes $p \geq 5$. The groups are Sylow pro-p-subgroups of the automorphism groups of certain semisimple $\mathbb{Q}_p$-Lie algebras. Details about the groups, e.g. generators of a Sylow pro-p subgroup, are given in Chapter XI. The tables are organised as follows.

1.) Dimension of the Lie algebra

2.) Name for the Lie algebra (usually the classical name e.g. $sl_2(\mathbb{Q}_p)$ except for the Lie algebra of type G_2, which is denoted by g_2)

3.) Primes and information on fields, e.g. for ramified extensions the minimal polynomial of a generator where numbers refer to chapter IX or for unramified extensions the degree since the extensions are (up to isomorphism) unique (In the general case for $p \geq 5$ ramified means any ramified extension of the appropriate degree.)

4.) Structural invariants of the groups

   (i) The isomorphism types of the factors of the lower central series. Define the following sequence g_i. If $\gamma_i(G)/\gamma_{i+1}(G)$ is elementary abelian i.e. $\gamma_i(G)/\gamma_{i+1}(G) \cong C_p^n$ define $g_i := n$. Otherwise if $\gamma_i(G)/\gamma_{i+1}(G) \cong C_{q_1}^{n_1} \times \cdots \times C_{q_\alpha}^{n_\alpha}$ where $q_1, \ldots, q_\alpha$ are distinct powers of p then define $g_i := q_1^{n_1} \cdot \cdots \cdot q_\alpha^{n_\alpha}$. For example if $\gamma_1(G)/\gamma_2(G) \cong C_2^2$ then $g_1 = 2$. If $\gamma_1(G)/\gamma_2(G) \cong C_2^3 \times C_4$ then $g_1 = 2^3 \cdot 4^1$.
   The $\overline{}$ indicates a period of the sequence. If $p = 2$ or $p = 3$ this is proved by the p-th powering map. For $p \geq 5$ the results follow from Chapter V and VI. The semicolon before the i-th term indicates that $\gamma_j(G)$ is powerful for $j \geq i$.

   (ii) The obliquity, if it was calculated. Namely the sequence

$$o_i := log_p(|\gamma_{i+1}(G) : \mu_i(G)|)$$

   where $\mu_i(G)$ is defined to be the intersection of $\gamma_{i+1}(G)$ with the intersection of all normal subgroups N of P with $N \not\leq \gamma_{i+1}(G)$. The $\overline{}$ indicates that the sequence repeats periodically which is proved by calculating in a sufficiently big quotient of the group, see Chapter X e). If this was too expensive, e.g. for groups with rather 'oblique' normal subgroups and big central sections it happens that the sequence obviously runs into some pattern but it has not yet been proved. We believe that the last k numbers will repeat where k is the length of the period of the lower central sequence marked by $\overline{}$
   The groups of dimension 12 are rather wide in the first sections. For these examples we give the ultimate obliquity

$$u_i := log_p(|\gamma_{i+1}(G) : \widetilde{\mu_i}(G)|)$$

where

$$\widetilde{\mu_i}(G) = \bigcap\{N | N \lhd G, N \leq \gamma_k(G), N \nleq \gamma_{i+1}(G)\} \cap \gamma_{i+1}(G)$$

for a chosen $k \in \mathbb{N}$. Unfortunately, it was not possible to prove by calculations in a finite quotient that the ultimate obliquity will repeat periodically like the given patterns because it is not clear whether one is deep enough down in the group. Choosing $k = 5$ makes the computation practically manageable. We give u_i only for those $i \in \mathbb{N}$ where we have the moral certainty that $u_i = o_i$. There are a few remaining groups where we could not give any information about the obliquity since the sections $\gamma_i(G)/\gamma_{i+1}(G)$ are too wide.

(iii) Define, if o_i is calculated

$$m_i := min\{j | \gamma_j(G) \leq \mu_i(G)\} - (i+1)$$

or if u_i is calculated

$$\widetilde{m_i} := min\{j | \gamma_j(G) \leq \widetilde{\mu_i}(G)\} - (i+1)$$

i.e. $\gamma_{i+1+m_i}(G)$ is the largest term of the lower central series contained in $\mu_i(G)$. Periodicity is treated as in (ii).

a) Dimension 3

$sl_2(\mathbb{Q}_p)$, cf. XI.c)

$p = 2$

g_i	$4, 2; \overline{2, 1}$
o_i	$5, 3, \overline{3, 2}$
m_i	$3, 2, \overline{3, 2}$

$p = 3$

g_i	$2; \overline{1, 2}$
o_i	$0, \overline{0, 0}$
m_i	$0, \overline{0, 0}$

$p \geq 5$

g_i	$2; \overline{1, 2}$

$sl_1(\mathcal{K}_2(\mathbb{Q}_p))$, cf. XI. e)

$p = 2$

g_i	$4, 2; \overline{2, 1}$
o_i	$5, 3, \overline{3, 2}$
m_i	$3, 2, \overline{3, 2}$

$p = 3$

g_i	$2; \overline{1, 2}$
o_i	$0, \overline{0, 0}$
m_i	$0, \overline{0, 0}$

$p \geq 5$

g_i	$2; \overline{1, 2}$

b) Dimension 6

$sl_2(K)$ with $|K : \mathbb{Q}_p| = 2$, cf. XI.c)

$p = 2$

$x^2 + 2,\ x^2 - 6$

$g_i \quad 4,\ \ 4,2,1,2,1;\overline{2,1,2,1}$
$o_i \quad 7,11,9,8,9,8,\overline{9,8,9,8}$
$m_i \quad 3,\ \ 7,6,5,6,5,\overline{6,5,6,5}$

$x^2 + 6,\ x^2 - 2$

$g_i \quad 5,\ \ 3,2,1,2,1;\overline{2,1,2,1}$
$o_i \quad 14,11,9,8,9,8,\overline{9,8,9,8}$
$m_i \quad 8,\ \ 7,6,5,6,5,\overline{6,5,6,5}$

$x^2 + 2x + 2\ ,\ x^2 + 2x + 6$

$g_i \quad 4,3,2,1,1,1;\overline{2,1,2,1}$
$o_i \quad 9,8,6,5,4,3,\overline{5,4,4,3}$
$m_i \quad 6,6,5,4,3,2,\overline{4,3,3,2}$

unramified of degree 2

$g_i \quad 4,3,2,3;\overline{2,2,2}$
$o_i \quad 10,7,5,4,\overline{6,6,4}$
$m_i \quad 5,4,3,3,\overline{4,4,3}$

$p = 3$

$x^2 + 3$

$g_i \quad 2,1,2;\overline{1,2,1,2}$
$o_i \quad 0,0,0,\overline{0,0,0,0}$
$m_i \quad 0,0,0,\overline{0,0,0,0}$

$x^2 - 3$

$g_i \quad 2,1,2;\overline{1,2,1,2}$
$o_i \quad 0,0,0,\overline{0,0,0,0}$
$m_i \quad 0,0,0,\overline{0,0,0,0}$

unramified of degree 2

$g_i \quad 4;\overline{2,4}$
$o_i \quad 0,\overline{0,0}$
$m_i \quad 0,\overline{0,0}$

$p \geq 5$

ramified of degree 2

$g_i \quad \overline{2,1,2,1}$

unramified of degree 2

$g_i \quad \overline{4,2}$

$sl_2(\mathbb{Q}_2)^2$, cf. (II.6)

$p = 2$

$g_i \quad 5,\ \ 4,2,2;\overline{2,2,1,1}$
$o_i \quad 14,10,8,6,\overline{8,6,5,4}$
$m_i \quad 7,\ \ 6,5,4,\overline{7,6,5,4}$

$sl_1(\mathcal{K}_2(K))$ with $|K:\mathbb{Q}_p| = 2$, cf. XI.e)

$p = 2$

x^2+2 , x^2-6

$$\begin{array}{ll} g_i & 4,\ \ 4,2,1,2,1;\overline{2,1,2,1} \\ o_i & 7,11,9,8,9,8,\overline{9,8,9,8} \\ m_i & 3,\ \ 7,6,5,6,5,\overline{6,5,6,5} \end{array}$$

x^2+6 , x^2-2

$$\begin{array}{ll} g_i & 2^3\cdot 4^1,\ \ 3,2,1,2,1;\overline{2,1,2,1} \\ o_i & 14,11,9,8,9,8,\overline{9,8,9,8} \\ m_i & 8,\ \ 7,6,5,6,5,\overline{6,5,6,5} \end{array}$$

x^2+2x+2 , x^2+2x+6

$$\begin{array}{ll} g_i & 4,3,2,1,1,1;\overline{2,1,2,1} \\ o_i & 9,8,6,5,4,3,\overline{5,4,4,3} \\ m_i & 6,6,5,4,3,2,\overline{4,3,3,2} \end{array}$$

unramified of degree 2

$$\begin{array}{ll} g_i & 2^3\cdot 4^1,2,2,1,2;\overline{1,2,1,2} \\ o_i & 10,8,6,5,5,\overline{4,6,6,5} \\ m_i & 7,6,5,4,4,\overline{3,5,5,4} \end{array}$$

$p = 3$

x^2+3

$$\begin{array}{ll} g_i & 2,1,2;\overline{1,2,1,2} \\ o_i & 0,0,0,\overline{0,0,0,0} \\ m_i & 0,0,0,\overline{0,0,0,0} \end{array}$$

x^2-3

$$\begin{array}{ll} g_i & 2,1,2;\overline{1,2,1,2} \\ o_i & 0,0,0,\overline{0,0,0,0} \\ m_i & 0,0,0,\overline{0,0,0,0} \end{array}$$

unramified of degree 2

$$\begin{array}{ll} g_i & 4;\overline{2,4} \\ o_i & 0,\overline{0,0} \\ m_i & 0,\overline{0,0} \end{array}$$

$p \geq 5$

ramified of degree 2

$$\begin{array}{ll} g_i & \overline{2,1,2,1} \end{array}$$

unramified of degree 2

$$\begin{array}{ll} g_i & \overline{4,2} \end{array}$$

$sl_1(\mathcal{K}_2(\mathbb{Q}_2))^2$, cf. (II.6)

$p = 2$

$$\begin{array}{ll} g_i & 5,\ \ 4,2,2;\overline{2,2,1,1} \\ o_i & 14,10,8,6,\overline{8,6,5,4} \\ m_i & 7,\ \ 6,5,4,\overline{7,6,5,4} \end{array}$$

c) Dimension 8

$sl_3(\mathbb{Q}_p)$, cf. XI.c)

$p = 2$

$$\begin{array}{ll} g_i & 3,2,2,1,2,1,2,2;\overline{1,2,1,2,2} \\ o_i & 3,1,3,2,4,3,1,3,\overline{2,4,3,1,3} \\ m_i & 2,1,2,1,3,2,1,2,\overline{1,3,2,1,2} \end{array}$$

$p = 3$

$$\begin{array}{ll} g_i & 3,1,2,1;\overline{1,1,2,1,2,1} \\ o_i & 5,4,2,1,\overline{0,0,5,4,2,1} \\ m_i & 4,3,2,1,\overline{0,0,4,3,2,1} \end{array}$$

$p \geq 5$

$$\begin{array}{ll} g_i & \overline{3,3,2} \end{array}$$

$su_3(K, \mathbb{Q}_p)$ with $|K : \mathbb{Q}_p| = 2$, cf. XI.d)

$p = 2$

$x^2 + 2$, $x^2 + 6$,
$x^2 - 2$, $x^2 + 6$

$$\begin{array}{ll} g_i & 5,\ 4,\ 4,\ 2,\ 3,\ 2;\ 3,\ \overline{2,\ 2,\ 2,\ 2} \\ o_i & 13,15,15,13,14,16,14,19,17,16,14 \\ m_i & 5,\ 6,\ 8,\ 7,\ 7,\ 9,\ 9,11,10,10,\ 9 \end{array}$$

$x^2 + 2x + 2$, $x^2 + 2x + 6$

$$\begin{array}{ll} g_i & 4,\ 4,\ 4,\ 2,\ 2,\ 2;\ \overline{2,\ 2,\ 2,\ 2} \\ o_i & 9,10,\ 9,14,12,11,\ 9,14,12,11 \\ m_i & 3,\ 5,\ 5,\ 7,\ 6,\ 6,\ 5,\ 7,\ 6,\ 6 \end{array}$$

unramified of degree 2

$$\begin{array}{ll} g_i & 3,2,2,1,2,1,2,2;\overline{1,2,1,2,2} \\ o_i & 3,1,3,2,4,3,1,3,2,4,3,1,3 \\ m_i & 2,1,2,1,3,2,1,2,1,3,2,1,2 \end{array}$$

$p = 3$

$x^2 + 3$

$$\begin{array}{ll} g_i & 4,2,2;\overline{1,2,2,3} \\ o_i & 10,8,6,\overline{5,4,2,6} \\ m_i & 6,5,4,\overline{3,2,1,4} \end{array}$$

$x^2 - 3$

$$\begin{array}{ll} g_i & 3,2,3;\overline{1,2,2,3} \\ o_i & 4,2,6,\overline{5,4,2,6} \\ m_i & 2,1,4,\overline{3,2,1,4} \end{array}$$

unramified of degree 2

$$\begin{array}{ll} g_i & 4,3;\overline{2,3,3} \\ o_i & 5,3,\overline{8,5,3} \\ m_i & 3,2,\overline{4,3,2} \end{array}$$

$p \geq 5$

unramified of degree 2

$$\begin{array}{ll} g_i & \overline{3,3,2} \end{array}$$

ramified of degree 2

$$\begin{array}{ll} g_i & \overline{2,1,1,1,2,1} \end{array}$$

$sl_1(\mathcal{K}_3(\mathbb{Q}_p))$, cf. XI.e)

$p = 2$

g_i	$3, 3, 2, 3, 3; \overline{2, 3, 3}$
o_i	$0, 0, 0, 3, 0, \overline{0, 3, 0}$
m_i	$0, 0, 0, 1, 0, \overline{0, 1, 0}$

$p = 3$

g_i	$3, 1, 2, 1; \overline{1, 1, 2, 1, 2, 1}$
o_i	$5, 4, 2, 1, \overline{0, 0, 5, 4, 2, 1}$
m_i	$4, 3, 2, 1, \overline{0, 0, 4, 3, 2, 1}$

$p \geq 5$

g_i	$\overline{3, 3, 2}$

d) Dimension 9

$sl_2(\mathbb{Q}_3)^3$, cf. (II.6)

$p = 3$

g_i	$3, 2, 2; \overline{1, 1, 1, 2, 2, 2}$
o_i	$4, 2, 0, \overline{0, 0, 0, 4, 2, 0}$
m_i	$2, 1, 0, \overline{0, 0, 0, 2, 1, 0}$

$sl_1(\mathcal{K}_2(\mathbb{Q}_3))^3$, cf. (II.6)

$p = 3$

g_i	$3, 2, 2; \overline{1, 1, 1, 2, 2, 2}$
o_i	$4, 2, 0, \overline{0, 0, 0, 4, 2, 0}$
m_i	$2, 1, 0, \overline{0, 0, 0, 2, 1, 0}$

$sl_2(K)$ with $|K : \mathbb{Q}_p| = 3$, cf. XI.c)

$p = 2$

$x^3 - 2$

g_i	$6, 2, 2, 2, 1, 1, 2, 1, 2, 1; \overline{2, 1, 2, 1, 2, 1}$
o_i	$19, 17, 15, 13, 12, 11, 12, 11, 12, 11, 12, 11, 12, 11, 12, 11$
m_i	$12, 11, 10, 9, 8, 7, 8, 7, 8, 7, 8, 7, 8, 7, 8, 7$

unramified

g_i	$8, 3, 5; \overline{3, 6}$

o_i and m_i unknown

$p = 3$

$x^3+3, x^3+12, x^3-6,$
$x^3+3x^2+3, x^3-3x+3,$
x^3+3x^2+3x+3
(non Galois extensions)

g_i $2,1,2,1,2;\overline{1,2,1,2,1,2}$
o_i $0,0,0,0,0,\overline{0,0,0,0,0,0}$
m_i $0,0,0,0,0,\overline{0,0,0,0,0,0}$

$x^3+3x-3, x^3+3x-12,$
x^3+3x+6
(Galois extensions)

g_i $3,1,2,1,2;\overline{1,2,1,2,1,2}$
o_i $3,0,0,0,0,\overline{0,0,0,0,0,0}$
m_i $1,0,0,0,0,\overline{0,0,0,0,0,0}$

unramified of degree 3

g_i $3,2,2;\overline{1,1,1,2,2,2}$
o_i $4,2,0,\overline{0,0,0,4,2,0}$
m_i $2,1,0,\overline{0,0,0,2,1,0}$

$p \geq 5$

unramified of degree 3

g_i $\overline{6,3}$

ramified of degree 3

g_i $\overline{2,1,2,1,2,1}$

$sl_1(\mathcal{K}_2(\mathbb{Q}_p))$ with $|K:\mathbb{Q}_p| = 3$, cf. XI.e)

$p = 2$

$x^3 - 2$

g_i	6,	2,	2,	2,	1,	1,	2,	1,	2,	1;	$\overline{2,}$	$\overline{1,}$	$\overline{2,}$	$\overline{1,}$	$\overline{2,}$	$\overline{1}$
o_i	19,	17,	15,	13,	12,	11,	12,	11,	12,	11,	12,	11,	12,	11,	12,	11
m_i	12,	11,	10,	9,	8,	7,	8,	7,	8,	7,	8,	7,	8,	7,	8,	7

unramified of degree 3

g_i $8,3,5;\overline{3,6}$
o_i and m_i unknown

$p = 3$

$x^3+3, x^3+12, x^3-6,$
$x^3+3x^2+3, x^3-3x+3,$
x^3+3x^2+3x+3
(not Galois extension)

g_i $2,1,2,1,2;\overline{1,2,1,2,1,2}$
o_i $0,0,0,0,0,\overline{0,0,0,0,0,0}$
m_i $0,0,0,0,0,\overline{0,0,0,0,0,0}$

$x^3+3x-3, x^3+3x-12,$
x^3+3x+6
(Galois extension)

g_i $3,1,2,1,2;\overline{1,2,1,2,1,2}$
o_i $3,0,0,0,0,\overline{0,0,0,0,0,0}$
m_i $1,0,0,0,0,\overline{0,0,0,0,0,0}$

unramified of degree 3

$$\begin{array}{ll} g_i & 3,2,2;\overline{1,1,1,2,2,2} \\ o_i & 4,2,0,\overline{0,0,0,4,2,0} \\ m_i & 2,1,0,\overline{0,0,0,2,1,0} \end{array}$$

$p \geq 5$

unramified of degree 3

$$\begin{array}{ll} g_i & \overline{6,3} \end{array}$$

ramified of degree 3

$$\begin{array}{ll} g_i & \overline{2,1,2,1,2,1} \end{array}$$

e) Dimension 10

$so_5(\mathbb{Q}_p) \cong sp_4(\mathbb{Q}_p)$ (split), cf. XI.f)

$p = 2$

$$\begin{array}{ll} g_i & 6,\ 4,\ 3,\ 3,\ 3;\ 2,\ 4,\ 3,\ 2,\ 3,\ 3,\ 3,\ \overline{3,\ 2,\ 3,\ 2} \\ o_i & 23,19,17,14,13,11,13,13,12,11,12,12,\overline{12,12,14,12} \\ m_i & 9,\ 8,\ 7,\ 6,\ 5,\ 4,\ 7,\ 6,\ 5,\ 6,\ 7,\ 6,\ \overline{7,\ 6,\ 7,\ 6} \end{array}$$

$p = 3$

$$\begin{array}{ll} g_i & 3,2,3;\overline{2,3,2,3} \\ o_i & 4,2,4,\overline{3,4,2,4} \\ m_i & 2,1,2,\overline{1,2,1,2} \end{array}$$

$p \geq 5$

$$\begin{array}{ll} g_i & \overline{3,2,3,2} \end{array}$$

$so_5(\mathbb{Q}_p) \cong sp_2(K_2(\mathbb{Q}_p))$ (non-split), cf. XI.g)

$p = 2$

$$\begin{array}{ll} g_i & 6,\ 4,\ 3,\ 3,\ 3,\ 2,\ 4,\ 3,\ 2,\ 3,\ 3,\ 3;\ \overline{3,\ 2,\ 3,\ 2} \\ o_i & 23,19,16,13,12,10,12,12,11,\ 9,12,11,10,12,13,11 \\ m_i & 9,\ 8,\ 7,\ 6,\ 5,\ 4,\ 7,\ 6,\ 5,\ 5,\ 7,\ 6,\ 6,\ 6,\ 7,\ 6 \end{array}$$

$p = 3$

$$\begin{array}{ll} g_i & 3,2,3;\overline{2,3,2,3} \\ o_i & 4,2,1,\overline{1,4,2,1} \\ m_i & 2,1,1,\overline{1,2,1,1} \end{array}$$

$p \geq 5$

$$\begin{array}{ll} g_i & \overline{3,2,3,2} \end{array}$$

f) Dimension 12

$sl_2(K)$ with $|K : \mathbb{Q}_p| = 4$, cf. XI.c)

$p = 2$

polynomials: 1-4,7-10,13-20,27-30

g_i $5, 5, 2, 2, 2, 1, 1, 1, 2, 1, 2, 1, 2, 1; \overline{2, 1, 2, 1, 2, 1, 2, 1}$

u_i $15, 14, 15, 14, 15, 14, 15, 14, 15, 14, 15, 14$

$\widetilde{m}_i$ $10, 9, 10, 9, 10, 9, 10, 9, 10, 9, 10, 9$

polynomials: 5,6,11,12,21-25,31

g_i $6, 4, 2, 2, 2, 1, 1, 1, 2, 1, 2, 1, 2, 1; \overline{2, 1, 2, 1, 2, 1, 2, 1}$

u_i $15, 14, 15, 14, 15, 14, 15, 14, 15, 14, 15, 14$

$\widetilde{m}_i$ $10, 9, 10, 9, 10, 9, 10, 9, 10, 9, 10, 9$

polynomials: 26,32

g_i $6, 3, 2, 1, 2, 1, 2, 1, 1, 1, 2, 1, 2, 1; \overline{2, 1, 2, 1, 2, 1, 2, 1}$

u_i $11, 10, 11, 10, 11, 10, 11, 10, 11, 10, 11, 10$

$\widetilde{m}_i$ $8, 7, 8, 7, 8, 7, 8, 7, 8, 7, 8, 7$

polynomials: 33,35

g_i $2^3 \cdot 4^1, 4, 3, 3, 2, 1, 1, 1, 2, 1, 2, 1, 2, 1; \overline{2, 1, 2, 1, 2, 1, 2, 1}$

u_i $15, 14, 15, 14, 15, 14, 15, 14, 15, 14, 15, 14$

$\widetilde{m}_i$ $10, 9, 10, 9, 10, 9, 10, 9, 10, 9, 10, 9$

polynomials: 34,36,37,39

g_i $2^4 \cdot 4^1, 3, 3, 3, 2, 1, 1, 1, 2, 1, 2, 1, 2, 1; \overline{2, 1, 2, 1, 2, 1, 2, 1}$

u_i $15, 14, 15, 14, 15, 14, 15, 14, 15, 14, 15, 14$

$\widetilde{m}_i$ $10, 9, 10, 9, 10, 9, 10, 9, 10, 9, 10, 9$

polynomials: 38,40

g_i $2^3 \cdot 4^1, 4, 3, 3, 2, 1, 1, 1, 2, 1, 2, 1, 2, 1; \overline{2, 1, 2, 1, 2, 1, 2, 1}$

u_i $15, 14, 15, 14, 15, 14, 15, 14, 15, 14, 15, 14$

$\widetilde{m}_i$ $10, 9, 10, 9, 10, 9, 10, 9, 10, 9, 10, 9$

polynomials: 41,43

g_i $6, 4, 2, 1, 2, 1, 2, 1, 1, 1, 2, 1, 2, 1; \overline{2, 1, 2, 1, 2, 1, 2, 1}$

u_i $10, 9, 11, 10, 10, 9, 11, 10, 10, 9, 11, 10$

$\widetilde{m}_i$ $8, 7, 8, 7, 8, 7, 8, 7, 8, 7, 8, 7$

polynomials: 42,44

g_i $6, 3, 3, 1, 2, 1, 2, 1, 1, 1, 2, 1, 2, 1; \overline{2, 1, 2, 1, 2, 1, 2, 1}$

u_i $10, 9, 11, 10, 10, 9, 11, 10, 10, 9, 11, 10$

$\widetilde{m}_i$ $8, 7, 8, 7, 8, 7, 8, 7, 8, 7, 8, 7$

polynomials: 45,46,47,48

g_i $7, 2, 2, 2, 2, 1, 1, 1, 2, 1, 2, 1, 2, 1; \overline{2, 1, 2, 1, 2, 1, 2, 1}$

u_i $15, 14, 15, 14, 15, 14, 15, 14, 15, 14, 15, 14$

$\widetilde{m}_i$ $10, 9, 10, 9, 10, 9, 10, 9, 10, 9, 10, 9$

polynomials: 49,50

g_i $6, 6, 4, 2, 2, 2; \overline{4, 2, 4, 2}$

u_i and $\widetilde{m}_i$ unknown

polynomials: 51,52

g_i $6, 5, 4, 2, 3, 2; \overline{4, 2, 4, 2}$

u_i and $\widetilde{m}_i$ unknown

polynomial: 53

g_i	$2^3 \cdot 4^1, 4, 3, 4, 2, 2, 2, 2, 2, 2; \overline{2, 2, 2, 2, 2, 2}$
u_i	$12, 10, 12, 12, 10, 12, 12, 10, 12, 12, 10$
$\widetilde{m}_i$	$7, 6, 7, 7, 6, 7, 7, 6, 7, 7, 6$

polynomial: 54

g_i	$2^3 \cdot 4^1, 4, 2, 3, 3, 1, 2, 2; \overline{1, 2, 3, 1, 2, 3}$
u_i	$10, 9, 10, 10, 9, 10, 10$
$\widetilde{m}_i$	$6, 5, 6, 6, 5, 5, 6$

polynomial: 55

g_i	$2^4 \cdot 4^1, 3, 3, 4, 2, 2, 2, 2, 2, 2; \overline{2, 2, 2, 2, 2, 2}$
u_i	$10, 12, 12, 10, 12, 12, 10, 12, 12, 10, 12$
$\widetilde{m}_i$	$7, 6, 7, 7, 6, 7, 7, 6, 7, 7, 6$

polynomials: 56,57

g_i	$6, 4, 3, 3, 2, 2, 2, 2, 2, 2; \overline{2, 2, 2, 2, 2, 2}$
u_i	$12, 10, 12, 12, 10, 12, 12, 10, 12, 12, 10$
$\widetilde{m}_i$	$7, 6, 7, 7, 6, 7, 7, 6, 7, 7, 6$

polynomial: 58

g_i	$6, 5, 2, 3, 2, 2, 2, 2, 2, 2; \overline{2, 2, 2, 2, 2, 2}$
u_i	$12, 10, 11, 12, 10, 11, 12, 10, 11, 12, 10$
$\widetilde{m}_i$	$7, 6, 7, 7, 6, 6, 7, 6, 7, 7, 6$

unramified of degree 4

g_i	$2^3 \cdot 4^1, 3, 3, 3, 2, 3, 3, 3; \overline{2, 2, 3, 3, 2}$
u_i	$8, 12, 11, 9, 10, 8$
$\widetilde{m}_i$	$5, 6, 5, 5, 6, 5$

$p = 3$

$x^4 + 3$, $x^4 - 3$

g_i	$2, 1, 2, 1, 2, 1, 2; \overline{1, 2, 1, 2, 1, 2, 1, 2}$
o_i	$0, 0, 0, 0, 0, 0, 0, \overline{0, 0, 0, 0, 0, 0, 0, 0}$
m_i	$0, 0, 0, 0, 0, 0, 0, \overline{0, 0, 0, 0, 0, 0, 0, 0}$

$x^2 + 3$, $x^2 - 3 \in K[x]$
with $K/\mathbb{Q}_p$ unramified of degree 2

g_i	$4, 2, 4; \overline{2, 4, 2, 4}$
o_i	$0, 0, 0, 0, \overline{0, 0, 0, 0}$
m_i	$0, 0, 0, 0, \overline{0, 0, 0, 0}$

unramified of degree 4

g_i $8; \overline{4, 8}$

o_i and m_i unknown

$p \geq 5$

totally ramified of degree 4

g_i $\overline{2, 1, 2, 1, 2, 1, 2, 1}$

degree 4, ramified of degree 2

g_i $\overline{4, 2, 4, 2}$

unramified of degree 4

g_i $\overline{8, 4}$

$sl_1(\mathcal{K}_2(K))$ with $|K : \mathbb{Q}_p| = 4$

$p = 2$

polynomials: 1-4,7-10,13-20,27-30

g_i $5, 5, 2, 2, 2, 1, 1, 1, 2, 1, 2, 1, 2, 1; \overline{2, 1, 2, 1, 2, 1, 2, 1}$

u_i $15, 14, 15, 14, 15, 14, 15, 14, 15, 14, 15, 14$

$\widetilde{m}_i$ $10, 9, 10, 9, 10, 9, 10, 9, 10, 9, 10, 9$

polynomials: 5,6,11,12,21-25,31

g_i $2^4 \cdot 4^1, 4, 2, 2, 2, 1, 1, 1, 2, 1, 2, 1, 2, 1; \overline{2, 1, 2, 1, 2, 1, 2, 1}$

u_i $15, 14, 15, 14, 15, 14, 15, 14, 15, 14, 15, 14$

$\widetilde{m}_i$ $10, 9, 10, 9, 10, 9, 10, 9, 10, 9, 10, 9$

polynomials: 26,32

g_i $2^4 \cdot 4^1, 3, 2, 1, 2, 1, 2, 1, 1, 1, 2, 1, 2, 1; \overline{2, 1, 2, 1, 2, 1, 2, 1}$

u_i $11, 10, 11, 10, 11, 10, 11, 10, 11, 10, 11, 10$

$\widetilde{m}_i$ $8, 7, 8, 7, 8, 7, 8, 7, 8, 7, 8, 7$

polynomial: 33

g_i $2^3 \cdot 4^1, 4, 3, 3, 2, 1, 1, 1, 2, 1, 2, 1, 2, 1; \overline{2, 1, 2, 1, 2, 1, 2, 1}$

u_i $15, 14, 15, 14, 15, 14, 15, 14, 15, 14, 15, 14$

$\widetilde{m}_i$ $10, 9, 10, 9, 10, 9, 10, 9, 10, 9, 10, 9$

polynomial: 34

g_i $2^3 \cdot 8^1, 3, 3, 3, 2, 1, 1, 1, 2, 1, 2, 1, 2, 1; \overline{2, 1, 2, 1, 2, 1, 2, 1}$

u_i $15, 14, 15, 14, 15, 14, 15, 14, 15, 14, 15, 14$

$\widetilde{m}_i$ $10, 9, 10, 9, 10, 9, 10, 9, 10, 9, 10, 9$

polynomials: 35,38,40

g_i $2^3 \cdot 4^1, 4, 3, 3, 2, 1, 1, 1, 2, 1, 2, 1, 2, 1; \overline{2, 1, 2, 1, 2, 1, 2, 1}$

u_i $15, 14, 15, 14, 15, 14, 15, 14, 15, 14, 15, 14$

$\widetilde{m}_i$ $10, 9, 10, 9, 10, 9, 10, 9, 10, 9, 10, 9$

polynomials: 36,37,39

g_i $2^3 \cdot 8^1, 3, 3, 3, 2, 1, 1, 1, 2, 1, 2, 1, 2, 1; \overline{2, 1, 2, 1, 2, 1, 2, 1}$

u_i $15, 14, 15, 14, 15, 14, 15, 14, 15, 14, 15, 14$

$\widetilde{m}_i$ $10, 9, 10, 9, 10, 9, 10, 9, 10, 9, 10, 9$

polynomials: 41-44

g_i $2^3 \cdot 4^1, 4, 3, 1, 2, 1, 2, 1, 1, 1, 2, 1, 2, 1; \overline{2, 1, 2, 1, 2, 1, 2, 1}$

u_i $10, 9, 11, 10, 10, 9, 11, 10, 10, 9, 11, 10$

$\widetilde{m}_i$ $8, 7, 8, 7, 8, 7, 8, 7, 8, 7, 8, 7$

polynomials: 45,46,47,48

g_i $7, 2, 2, 2, 2, 1, 1, 1, 2, 1, 2, 1, 2, 1; \overline{2, 1, 2, 1, 2, 1, 2, 1}$

u_i $15, 14, 15, 14, 15, 14, 15, 14, 15, 14, 15, 14$

$\widetilde{m}_i$ $10, 9, 10, 9, 10, 9, 10, 9, 10, 9, 10, 9$

polynomials: 49,50

g_i $6, 6, 4, 2, 2, 2; \overline{4, 2, 4, 2}$

u_i and $\widetilde{m}_i$ unknown

polynomials: 51,52

g_i $6, 5, 4, 2, 3, 2; \overline{4, 2, 4, 2}$

u_i and $\widetilde{m}_i$ unknown

polynomial: 53

$g_i \quad 2^3\cdot 4^1,3,3,2,2,2,2,\ 1,\ 1,\ 1,\ 2,\ 2,\ 1;\ \overline{1,\ 2,\ 2,\ 1,\ 1,\ 2,\ 2,\ 1}$

$u_i \quad 12,11,10,12,12,11,10,12,12,11,10,12,12,11$

$\widetilde{m}_i \quad 9,\ 8,\ 7,\ 9,\ 9,\ 8,\ 7,\ 9,\ 9,\ 8,\ 7,\ 9,\ 9,\ 8$

polynomial: 54

$g_i \quad 2^3\cdot 4^1,3,2,2,2,2,1,1,2,\ 1,\ 1,\ 1;\ \overline{2,\ 2,\ 1,\ 1,\ 2,\ 2,\ 1,\ 1}$

$u_i \quad 11,10,\ 9,10,11,10,\ 9,10,11,10,\ 9$

$\widetilde{m}_i \quad 8,\ 7,\ 6,\ 8,\ 8,\ 7,\ 6,\ 7,\ 8,\ 7,\ 6$

polynomial: 55

$g_i \quad 2^3\cdot 8^1,2,3,2,2,2,2,\ 1,\ 1,\ 1,\ 2,\ 2,\ 1;\ \overline{1,\ 2,\ 2,\ 1,\ 1,\ 2,\ 2,\ 1}$

$u_i \quad 12,11,10,12,12,11,10,12,12,11,10,12,12,11$

$\widetilde{m}_i \quad 9,\ 8,\ 7,\ 9,\ 9,\ 8,\ 7,\ 9,\ 9,\ 8,\ 7,\ 9,\ 9,\ 8$

polynomial: 56

$g_i \quad 2^3\cdot 4^1,4,3,1,2,2,2,\ 1,\ 1,\ 1,\ 2,\ 2,\ 1;\ \overline{1,\ 2,\ 2,\ 1,\ 1,\ 2,\ 2,\ 1}$

$u_i \quad 12,11,10,12,12,11,10,12,12,11,10,12,12,11$

$\widetilde{m}_i \quad 9,\ 8,\ 7,\ 9,\ 9,\ 8,\ 7,\ 9,\ 9,\ 8,\ 7,\ 9,\ 9,\ 8$

polynomial: 57

$g_i \quad 2^2\cdot 4^2,3,3,1,2,2,2,\ 1,\ 1,\ 1,\ 2,\ 2,\ 1;\ \overline{1,\ 2,\ 2,\ 1,\ 1,\ 2,\ 2,\ 1}$

$u_i \quad 12,11,10,12,12,11,10,12,12,11,10,12,12,11$

$\widetilde{m}_i \quad 9,\ 8,\ 7,\ 9,\ 9,\ 8,\ 7,\ 9,\ 9,\ 8,\ 7,\ 9,\ 9,\ 8$

polynomial: 58

$g_i \quad 2^3\cdot 4^1,4,2,2,2,2,2,\ 1,\ 1,\ 1,\ 2,\ 2,\ 1;\ \overline{1,\ 2,\ 2,\ 1,\ 1,\ 2,\ 2,\ 1}$

$u_i \quad 12,11,10,11,12,11,10,11,12,11,10,11,12,11$

$\widetilde{m}_i \quad 9,\ 8,\ 7,\ 9,\ 9,\ 8,\ 7,\ 8,\ 9,\ 8,\ 7,\ 9,\ 9,\ 8$

unramified of degree 4

$g_i \quad 2^3\cdot 8^1,2,2,2,1,1,2,1,2,2,\ 2;\ \overline{1,\ 1,\ 1,\ 2,\ 2,\ 2,\ 2,\ 1}$

$u_i \quad 9,\ 8,\ 7,\ 6,12,11,\ 9,10,\ 9$

$\widetilde{m}_i \quad 8,\ 7,\ 6,\ 5,\ 9,\ 8,\ 7,\ 9,\ 8$

$p = 3$

$x^4+3,\ x^4-3$

$g_i \quad 2,1,2,1,2,1,2;\overline{1,2,1,2,1,2,1,2}$

$o_i \quad 0,0,0,0,0,0,0,\overline{0,0,0,0,0,0,0,0}$

$m_i \quad 0,0,0,0,0,0,0,\overline{0,0,0,0,0,0,0,0}$

$x^2+3,\ x^2-3 \in K[x]$
with $K/\mathbb{Q}_p$ unramified of degree 2

$g_i \quad 4,2,4;\overline{2,4,2,4}$

$o_i \quad 0,0,0,\overline{0,0,0,0}$

$m_i \quad 0,0,0,\overline{0,0,0,0}$

unramified of degree 4

$g_i \quad 8;\overline{4,8}$

o_i and m_i unknown

$p \geq 5$

totally ramified of degree 4

$g_i \quad \overline{2,1,2,1,2,1,2,1}$

degree 4, ramified of degree 2

$g_i \quad \overline{4,2,4,2}$

unramified of degree 4

$$g_i \quad \overline{8,4}$$

$sl_2(K)^2$ with $|K : \mathbb{Q}_p| = 2$, cf. (II.6)

$p = 2$

$x^2 + 2, x^2 - 6$

$$\begin{array}{lll} g_i & 5,4,4,4,2,2,1,1, & 2,\ 2,\ 1,\ 1;\ \overline{2,\ 2,\ 1,\ 1,\ 2,\ 2,\ 1,\ 1} \\ u_i & & 20,18,17,16,20,18,17,16,20,18,17,16 \\ \widetilde{m}_i & & 13,12,11,10,13,12,11,10,13,12,11,10 \end{array}$$

$x^2 - 2, x^2 + 6$

$$\begin{array}{lll} g_i & 6,5,3,3,2,2,1,1,2,2,1,1; & \overline{2,\ 2,\ 1,\ 1,\ 2,\ 2,\ 1,\ 1} \\ u_i & & 20,18,17,16,20,18,17,16 \\ \widetilde{m}_i & & 13,12,11,10,13,12,11,10 \end{array}$$

$x^2 + 2x + 2, x^2 + 2x + 6$

$$\begin{array}{lll} g_i & 5,4,3,3,2,2,1,1,1,1,1,1; & \overline{2,\ 2,\ 1,\ 1,\ 2,\ 2,\ 1,\ 1} \\ u_i & & 12,10,\ 9,\ 8,10,\ 8,\ 7,\ 6 \\ \widetilde{m}_i & & 9,\ 8,\ 7,\ 6,\ 7,\ 6,\ 5,\ 4 \end{array}$$

unramified of degree 2

$$\begin{array}{lll} g_i & 6,5,3,3,2,2,3,3; & \overline{2,\ 2,\ 2,\ 2,\ 2,\ 2} \\ u_i & & 14,12,14,12,10,\ 8 \\ \widetilde{m}_i & & 10,\ 9,10,\ 9,\ 7,\ 6 \end{array}$$

$sl_1(\mathcal{K}_2(K))^2$ with $|K : \mathbb{Q}_p| = 2$, cf. (II.6)

$p = 2$

$x^2 + 2, x^2 - 6$

$$\begin{array}{lll} g_i & 5,4,4,4,2,2,1,1,2,2,1,1; & \overline{2,\ 2,\ 1,\ 1,\ 2,\ 2,\ 1,\ 1} \\ u_i & & 20,18,17,16,20,18,17,16 \\ \widetilde{m}_i & & 13,12,11,10,13,12,11,10 \end{array}$$

$x^2 - 2, x^2 + 6$

$$\begin{array}{lll} g_i & 2^4 \cdot 4^1,4,4,3,2,2,1,1,2,2,1,1; & \overline{2,\ 2,\ 1,\ 1,\ 2,\ 2,\ 1,\ 1} \\ u_i & & 20,18,17,16,20,18,17,16 \\ \widetilde{m}_i & & 13,12,11,10,13,12,11,10 \end{array}$$

$x^2 + 2x + 2, x^2 + 2x + 6$

$$\begin{array}{lll} g_i & 5,4,3,3,2,2,1,1,1,1,1,1; & \overline{2,\ 2,\ 1,\ 1,\ 2,\ 2,\ 1,\ 1} \\ u_i & & 12,10,\ 9,\ 8,10,\ 8,\ 7,\ 6 \\ \widetilde{m}_i & & 9,\ 8,\ 7,\ 6,\ 7,\ 6,\ 5,\ 4 \end{array}$$

unramified of degree 2

$$\begin{array}{lll}
g_i & 2^4\cdot 4^1,4,3,2,2,2,1,1,2,2,1,1; & \overline{2,\ 2,\ 2,\ 2,\ 1,\ 1,\ 1,\ 1} \\
u_i & & 14,12,14,12,11,10,\ 9,\ 8 \\
\widetilde{m}_i & & 11,10,11,10,\ 9,\ 8,\ 7,\ 6
\end{array}$$

$sl_2(\mathbb{Q}_2)^4$, cf. (II.6)

$p=2$

$$\begin{array}{lll}
g_i & 6,5,4,4,2,2,2,2; & \overline{2,\ 2,\ 2,\ 2,\ 1,\ 1,\ 1,\ 1} \\
u_i & & 18,16,14,12,11,10,\ 9,\ 8 \\
\widetilde{m}_i & & 15,14,13,12,11,10,\ 9,\ 8
\end{array}$$

$sl_1(\mathcal{K}_2(\mathbb{Q}_2))^4$, cf. (II.6)

$p=2$

$$\begin{array}{lll}
g_i & 6,5,4,4,2,2,2,2; & \overline{2,\ 2,\ 2,\ 2,\ 1,\ 1,\ 1,\ 1} \\
u_i & & 18,16,14,12,11,10,\ 9,\ 8 \\
\widetilde{m}_i & & 15,14,13,12,11,10,\ 9,\ 8
\end{array}$$

g) Dimension 14

$g_2(\mathbb{Q}_p)$ (simple Lie algebra of type G_2), cf. XI.h)

$p=2$

$$\begin{array}{ll}
g_i & 4,4,3,2,3,2,4,3;\overline{2,3,2,4,3} \\
o_i & 8,6,7,5,9,8,6,7,5,9,8,6,7 \\
m_i & 3,4,4,3,4,3,4,4,3,4,3,4,4
\end{array}$$

$p=3$

$$\begin{array}{lll}
g_i & 4,\ 3,\ 3,\ 2;\ 3,\ 3,\ 3,\ 2,\ 4,\ 3, & \overline{3,\ 2,\ 3,\ 2,\ 2,\ 2} \\
o_i & 11,12,10,\ 9,11,11,11,\ 9,11,10, & 12,10,14,12,10,10 \\
m_i & 4,\ 5,\ 4,\ 3,\ 4,\ 5,\ 6,\ 5,\ 6,\ 7, & 8,\ 7,\ 8,\ 7,\ 6,\ 7
\end{array}$$

$p\geq 5$

$$g_i \quad \overline{3,2,2,2,3,2}$$

XIII Uncountably many just infinite pro-p-groups of finite width

a) The Nottingham group

Let G_q be the group of continuous automorphisms of the ring $\mathbb{F}_q[[t]]$ where q is a prime power. Clearly G_q can also be regarded as the group of continuous automorphisms of $\mathbb{F}_q((t))$. In fact every automorphism of $\mathbb{F}_q[[t]]$ is continuous. The simplest proof is to observe that an element b of $\mathbb{F}_q[[t]]$ is of the form $a^q - a$ if and only if b has positive valuation. Thus every automorphism preserves the valuation on $\mathbb{F}_q[[t]]$, and hence is continuous. Similarly one can prove that every automorphism of $\mathbb{F}_q((t))$ is continuous, but we do not need these facts, since we are in any case only concerned with continuous automorphisms.
Clearly any element of G_q is determined by its action on t, and this can be any element $\sum_{i=1}^{\infty} a_i t^i$, where the a_i lie in $\mathbb{F}_q$, and $a_1 \neq 0$.

(XIII.1) Definition. *Let G_q be the group of continuous automorphisms of the ring $\mathbb{F}_q[[t]]$. The Nottingham group S_q is defined to be the Sylow pro-p-subgroup of G_q; that is, the subgroup of $G_q = \{\sum_{i=1}^{\infty} a_i t^i \mid a_i \in \mathbb{F}_q, a_1 \neq 0\}$ defined by the condition $a_1 = 1$.*

Suppose now that q is a prime $p > 2$. If we define e_i in S_p to be given by $t \mapsto t(1+t^i)$ then it is easy to see that $(e_1, e_2, \ldots)$ forms a base for S_p; that is, every element of S_p can be written uniquely as $e_1^{k_1} e_2^{k_2} \cdots$, where $0 \leq k_i < p$ for all i. If $S_p(i)$ is the subgroup of S_p that centralises $t \bmod (t^{i+1})$ then $S_p(i)$ is generated topologically by $\{e_i, e_{i+1}, \ldots\}$, and it is easy to see that $[e_i, e_j] \equiv e_{i+j}^{i-j} \bmod S_p(k)$ where $k = \min(2i+j, i+2j)$. From this it follows at once that S_p is just infinite, with width 2, and $\gamma_i(S_p)/\gamma_{i+1}(S_p)$ has order p^2 if $i \equiv 1 \bmod (p-1)$, and has order p otherwise. In fact $\gamma_i(S_p) = S_p(r)$ where $r = i + \lceil (i-1)/(p-1) \rceil$.
Similarly, it is easy to see that $S_p(i)^p \leq S_p(2i)$, so S_p is not p-adic analytic.
These results are easily generalised to q being a prime power. If q is an odd prime power, the i-th lower central factor of S_q may be naturally regarded as a vector space of dimension 1 or 2 over $\mathbb{F}_q$, the dimension being 2 exactly when $i \equiv 1 \bmod (p-1)$. For $q = 2$, $S_p/S_p(2) \cong C_2 \times C_4$, but the other terms of the lower central series are of dimension 1 or 2 over $\mathbb{F}_2$. The case when q is a power of 2 is similar. In all cases, S_q is a just infinite non-p-adic analytic group of finite width. For details see [Joh 88].

b) Construction of uncountably many groups

Let p be a prime. Note that the group of units U of $\mathbb{F}_p[[t]]$ is the direct product of a cyclic group of order $p-1$ with a free $\mathbb{Z}_p$-module U_1 on $\{l_i \mid p \nmid i\}$, where $l_i = 1 + t^i$. That is to say, given any $\mathbb{Z}_p$-module M containing a sequence $\{f_i \mid p \nmid i\}$ that converges to 0, there is a unique homomorphism of U_1 into M mapping l_i to f_i for all i (with $p \nmid i$). It follows that $V := U_1/U_1^p$ is a compact vector space over $\mathbb{F}_p$, with topological basis $\{v_i \mid p \nmid i\}$ with $v_i = U_1^p l_i$. The Nottingham group S_p acts faithfully on this vector space. The stabiliser of $v_i = U_1^p l_i$ is a closed subgroup $C_{(i)}$ of infinite order and index in S_p. Now V has a filtration $V = V_1 \geq V_2 \geq \cdots$, where $V_i = \langle v_j \mid j \geq i, p \not| j \rangle$. If $v \in V_i - V_{i+1}$, define $wt(v) = i$. If W is a finite dimensional subspace of V, then W has a basis $(b_1, \ldots, b_d)$ with $wt(b_i) < wt(b_{i+1})$ for $1 \leq i < d$. Then $(wt(b_1), \cdots, wt(b_d))$ is an invariant for the orbit of W under S_p.

(XIII.2) Lemma. *If $d(1 - 1/p) > 1$ (i.e. $d > 1$ and $p > 2$ or $d > 2$ and $p = 2$) and U_1 is as defined above, then the Nottingham group S_p acts on the set of finite dimensional subspaces of U_1/U_1^p, of given weight invariant $(w_1, ..., w_d)$, with uncountably many orbits.*

Proof. Let W be such a subspace. Now W has a unique reduced echelonised basis w.r.t. the above basis $\{v_i \mid p \not| i\}$, and the ith basis element can be written as $\sum a_j v_j$, with a_j in $\mathbb{F}_p$, where j runs through the positive integers prime to p, $a_j = 0$ if $j < w_i$, $a_j = 1$ if $j = w_i$, and for every $j > i$ $a_j = 0$ if $j = w_k$ for some $i < k \leq d$. There are no other restrictions on these basis vectors, so the set of subspaces W with the given invariant weights is in natural 1-1 correspondence with the set S of ordered d-tuples of sequences (a_j) satisfying these conditions. Now define an equivalence relation $\underset{n}{\sim}$ on S for each $n > 0$ so that two elements of S are n-equivalent if and only if the corresponding sequences agree on terms a_j for $j < n$. Since j cannot be a multiple of p, $\underset{kp}{\sim}$ and $\underset{kp+1}{\sim}$ are the same for $k > 0$. Now let $S_p(n)$ be the subgroup of S_p that fixes t modulo t^{n+1}. It is clear that S_p permutes these n-equivalence classes, and that $S_p(n)$ fixes them. But $S_p/S_p(n)$ is of order p^{n-1}, and the number of equivalence classes under $\underset{n}{\sim}$ is at least $p^{dn(1-1/p)-c}$ for some easily calculated constant c. It follows that the number of orbits of S_p on the n-equivalence classes increases exponentially with n since $d > 1$ and $p > 2$. q.e.d.

Define $(PSL_p(\mathbb{F}_p[[t]]))_p$ to be a Sylow pro-p-subgroup of $PSL_p(\mathbb{F}_p[[t]])$. Now let W be a d-dimensional subspace of V, and let $u_1, \cdots, u_d \in U_1$ define a basis for W. Let $P_W < PGL_p(\mathbb{F}_p[[t]])$ be generated by

$(PSL_p(\mathbb{F}_p[[t]]))_p$ and $\left\{ A_i = \begin{pmatrix} u_i & 0 \\ 0 & I_{p-1} \end{pmatrix} \mid 1 \leq i \leq d \right\}$ modulo scalars.

Now $A_i^p = u_i I_p \cdot \begin{pmatrix} u_i^{p-1} & 0 \\ 0 & u_i^{-1} I_{p-1} \end{pmatrix} \in (PSL_p(\mathbb{F}_p[[t]]))_p$ modulo scalars, so P_W is an extension of $(PSL_p(\mathbb{F}_p[[t]]))_p$ by an elementary abelian group of order p^d, and clearly P_W is determined by W. Conversely, det maps $PGL_p(\mathbb{F}_p[[t]])$ homomorphically onto U/U^p, and maps P_W onto W, so P_W determines W. Now assume that $p > 2$. Let W_1 and W_2 be finite dimensional subspaces of V. It is clear from [HOM 89] Proposition 3.2.7 that P_{W_1} and P_{W_2} are full groups (cf. [HOM 89] p. 106) since $p > 2$, so by [HOM 89] Theorem 3.2.29, any isomorphism between these groups is the restriction of an automorphism of $PGL_p(\mathbb{F}_p[[t]])$, and this automorphism group is generated by inner automorphisms, the inverse transpose automorphism, and field automorphisms. Only the latter affects the determinant, and hence the isomorphism exists if and only if W_1 and W_2 are in the same orbit of the Nottingham group. This gives us uncountably many finitely generated non-isomorphic groups of finite width. It is clear that, since U_1 contains no elements of order p, these groups are all just infinite.

XIV Some open problems

a) Problems on general just infinite groups

(i) Assume P is an insoluble hereditarily just infinite group satisfying $w(P) < \infty$. Is every open subgroup Q of P also of width $w(Q) < \infty$?

(ii) Is there a hereditarily just infinite pro-p-group of finite width such that the series $|\gamma_i(P) : \gamma_{i+1}(P)|$ for $i \in \mathbb{N}$ is not periodic?
Can one find such an example as an open subgroup of the Nottingham group?

(iii) Is every hereditarily just infinite pro-p-group of finite width finitely presented as a pro-p-group?

(iv) Is every just infinite pro-p-group of finite width also of finite obliquity? If so, it gives a positive answer to (i).

(v) Does average width 1 imply finite coclass?

(vi) Is it the case that for every just infinite pro-p-group P of finite width, there is a natural number i such that, if Q is any infinite pro-p-group with $Q/\gamma_i(Q) \cong P/\gamma_i(P)$ then Q is just infinite of finite width? Note that an affirmative answer would give an affirmative answer to (iii). Since there are uncountably many just infinite pro-p-groups of finite width, we cannot hope that for large enough i the isomorphism $Q/\gamma_i(Q) \cong P/\gamma_i(P)$ should imply $Q \cong P$, as in the case if P is a $\tilde{p}$-group.

b) Problems on $\tilde{p}$-groups

(i) Find a finite pro-p-presentation for every maximal $\tilde{p}$-group (either algorithmically or theoretically).

(ii) Implement an algorithm to determine, from a sufficiently large quotient of a $\tilde{p}$-group Q, the maximal $\tilde{p}$-group P containing Q as an open subgroup.

(iii) Can one use the Cayley version of the Baker-Campbell-Hausdorff-formula to extend some of the investigations of Chapter VI to the case $p = 2$?

XV References

Bla 58 N. Blackburn, *On a special class of p-groups.* Acta Math. **100** (1958), 45-92.

Bla 61 N. Blackburn, *Generalization of certain elementary theorems on p-groups.* Proc. London Math. Soc. **11** (1961), 1-22.

BrT 72 F. Bruhat, J. Tits, *Groupes réductives sur un corps local I. Donnée radicielles valuées.* Publ. Math. I. H. E. S. **41** (1972), 5-251.

BrT 84 F. Bruhat, J. Tits, *Groupes réductives sur un corps local II. Schéma en groupes. Existence d'une donnée radicille valuée.* Publ. Math. I. H. E. S. **60** (1984), 197-376.

BrT 87 F. Bruhat, J. Tits, *Groupes algébriques sur un corps local III. Complements et applications á cohomologie galoisienne.* J. Fac. Sci. Univ. Tokyo Sec. 1A **34** (1984), 671-688.

Cam 97 R. Camina, *Subgroups of the Nottingham group.* Journal of Algebra, to appear.

Car 72 R. W. Carter, *Simple groups of Lie type.* Wiley, London 1972.

Che 55 C. Chevalley, *Sur certains groupes simples.* Tohoku Math. J. (2) **7**, (1955), 14-66.

DdMS 91 J. D. Dixon, M. P. F. du Sautoy, A. Mann, D. Segal, *Analytic pro-p Groups.* LMS Lecture Note Series 157, 1991.

Don 87 S. Donkin, *Space Groups and Groups of Prime-Power Order VIII. Pro-p-Groups of Finite Coclass and p-Adic Lie Algebras.* J. Alg. **111** no. 2 (1987), 316-342.

GAP 94 M. Schönert (ed.), *Groups, Algorithms, and Programming.* GAP-3.4 Manual, Lehrstuhl D für Mathematik, RWTH Aachen.

Gri 80 R. I. Grigorchuk, *On the Burnside problem for periodic groups.* Funktsional. Anal. i Prilozhen. 14 (1980), 53 - 54, [Russian]. Engl. transl.: Functional Anal. Appl. 14 (1980), 41-43.

Has 49 H. Hasse, *Zahlentheorie.* Akademie-Verlag 1949, 3rd ed. 1969.

HOM 89 A. J. Hahn, O. T. O'Meara, *The Classical Groups and K-Theory.* Springer Berlin 1989.

Hup 67 B. Huppert, *Endliche Gruppen I.* Springer Berlin 1967.

Iwa 66 N. Iwahori, *Generalized Tits system (Bruhat decomposition) on p-adic semisimple groups* in *Algebraic groups and discontinuous Subgroups.* Proc. Sympos. Pure Math., vol. IX, Part 1, 71-83, Providence RI 1966.

Iws 86 K. Iwasawa, *Local Class Field Theory.* Oxford University Press 1986.

Jac 62 N. Jacobson, *Lie Algebras.* Wiley New York 1962.

Joh 88 D. L. Johnson, *The group of formal power series under substitution.* J. Austral. Math. Soc. (Series A) **45** (1988), 298-302.

Kne 65 M. Kneser, *Galois-Kohomologie halbeinfacher algebraischer Gruppen über p-adischen Körpern. I.*, Math. Z. **88** (1965), 40-47; II., ibid. **89** (1965), 250-272.

Kne 69 M. Kneser, *Lectures on Galois Cohomology of Classical Groups.* Tata Institute of Fundamental Research 1969.

Kra 62 M. Krasner, *Nombres des extensions d'un degrè donneè d'un corp p-adic.* Comptes Rendues Hebdomadaires, Academie des Sciences, Paris **254**, **255**, 1962.

Laz 65 M. Lazard, *Groupes analytiques p-adiques.* Inst. Hautes Études Scientifiques, Publ. Math. 26 (1965), 389-603.

LeN 80 C. R. Leedham-Green, M. F. Newman, *Space groups and groups of prime-power order I.* Arch. d. Math. **35** (1980), 193-202.

Lee 94a C. R. Leedham-Green, *Pro-p-groups of finite coclass.* J. London Math, Soc. **50** (1994), 43-48.

Lee 94b C. R. Leedham-Green, *The structure of finite p-groups.* J. London Math, Soc. **50** (1994), 49-67.

LuM 87a A. Lubotzky, A. Mann, *Powerful p-groups. I: finite groups.* J. Algebra **105** (1987), 484-506.

LuM 87b A. Lubotzky, A. Mann, *Powerful p-groups. II: p-adic analytic groups.* J. Algebra **105** (1987), 506-515.

MAG 95 W. Bosma, J. Cannon, *Handbook of* MAGMA *functions.* School of Mathematics and Statistics, Sydney University 1995.

Nar 90 W. Narkiewicz, *Elementary and Analytic Theory of Algebraic Numbers.* Springer Berlin 1990.

Neu 86 J. Neukirch, *Class Field Theory.* Springer Berlin 1986.

OMe 73 O. T. O'Meara, *Introduction to Quadratic Forms.* Springer Berlin 1973.

PlP 77 W. Plesken, M. Pohst, *On Maximal Irreducible Subgroups of $GL(n,\mathbb{Z})$ I. The Five and Seven Dimensional Cases.* Mathematics of Computation vol. 31, no. 138, (1977), 536-551.

Rei 75 I. Reiner, *Maximal Orders.* Academic Press London 1975.

Roz 96 A. V. Rozhkov, *Lower central series of a group of automorphisms.* Mathematical Notes of Sciences of USSR, 1996, vol. 60, no. 1-2, 165-174, (Engl. transl. of Matematiceskie Zametki).

Sat 71 I. Satake, *Classification Theory of Semi-Simple Algebraic Groups.* Lecture Notes in Pure and Applied Math., M. Dekker New York 1971.

Ser 92 J.-P. Serre, *Lie Algebras and Lie Groups.* Springer Berlin 1992.

Sha 94 A. Shalev, *The structure of finite p-groups: effective proof of the coclass conjectures.* Invent. Math. **115** (1994), 315-346.

Sid 84 S. Sidki, *On a 2-Generated Infinite 3-Group: Subgroups and Automorphisms.* Journal of Algebra **110**, (1987), 24-55.

Sou 96 B. Souvignier, *Erweiterungen von analytischen pro-p-Gruppen mit endlichen Gruppen.* PhD thesis, RWTH Aachen 1996.

Ste 61 R. Steinberg, *Automorphisms of classical Lie algebras.* Pacific J. Math., **11** (1961), 1119-1129.

Ste 68 R. Steinberg, *Lectures on Chevalley groups.* Mimeographed lecture notes, Yale University Mathematics Department, New Haven, CT, 1968.

Scha 74 W. Scharlau, *Involutions on orders. I.* Journ. reine u. angew. Math. **268/269**, 190-202 (1975).

Scha 85 W. Scharlau, *Quadratic and Hermitian forms.* Springer Berlin 1985.

Tay 92 D. E. Taylor, *The Geometry of the Classical Groups.* Heldermann Verlag Berlin 1992.

Tit 79 J. Tits, *Reductive groups over local fields.* Proc. Symp. Pure Math. vol. 33, part 1 (Corvallis 77), Amer. Math. Soc. (1979), 22-69.

Tsu 61 T. Tsukamoto, *On the local theory of quaternionic anti-hermitian forms.* J. Math. Soc. Japan, vol. 13, no. 4 (1961), 387-400.

Wei 61 A. Weil, *Algebras with involutions and the classical groups.* J. Ind. Math. Soc. **24** (1961), 589-623.

Wey 46 H. Weyl, *The Classical Groups, Their Invariants and Representation.* Princeton University Press 1946.

Zas 39 H. Zassenhaus, *Über Lie'sche Ringe mit Primzahlcharakteristik.* Abh. math. Sem. Univ. Hamburg **13** (1939), 1-100.

Authors' addresses and e-mail:

Queen Mary and Westfield College,
University of London
School of Mathematical Sciences
Mile End Road
London E1 4NS
England
C.R.Leedham-Green@qmw.ac.uk

RWTH Aachen
Lehrstuhl B für Mathematik
Templergraben 64
52062 Aachen
Germany
plesken@willi.math.rwth-aachen.de
gz@willi.math.rwth-aachen.de

XVI Notation

$\lfloor a \rfloor$	the greatest integer $\leq a$
$\lceil a \rceil$	the least integer $\geq a$
$\overline{lim}$	limes superior
$\wr$	wreath product
$\rtimes$	semi-direct product of
$Z(G)$	centre of the group G
$[X, X]$	commutator group if X is a group Lie commutator of X is a Lie algebra
G^p	$= \langle g^p \vert g \in G \rangle$ for a group G
$\gamma_i(G)$	the closure of the i-th term of the lower central series of G, i.e. $\gamma_1 := G, \gamma_{i+1} := \overline{[\gamma_i, G]}$ if G is a finite group or a finitely generated pro-p-group then the terms of the lower central series are closed automatically
$\lambda_i(G)$	the closure of the i-th term of the lower p-series of G, i.e. $\lambda_1 := G, \lambda_{i+1} := \overline{[\lambda_i, G]\lambda_i(G)^p}$ if G is a finite group or a finitely generated pro-p-group then the terms of the lower central series are closed automatically
ν, ν_p	valuation, valuation where p is of value 1
$\mathcal{O}_K$	ring of integers of field K
$\mathcal{O}$	ring of integers of some field
$\mathcal{O}_i^*$	$= \{x \mid \nu(x-1) \geq i\}$
Π or π	prime elements
F	residue class field
ζ_k	primitive k^{th} root of unity
$w(P)$	width of P (I.1)
$w_a(P)$	average width of P (I.1)
$\overline{w}(P)$	ultimate width of P (I.1)
$\overline{w_a}(P)$	upper average width of P (I.1)
$o(P)$	obliquity of P (I.5)
$o_i(P)$	i-th obliquity of P (I.5)
μ_i	(I.5)
$\tilde{p}$-group	p-adic analytic just infinite pro-p group (I.5)
$[H,_k G]$	k-fold commutator cf. (II.1)
$\Phi(x, y)$	Baker-Campbell-Hausdorff formula, cf. Chapter III a)
$\Psi(x, y)$	commutator Baker-Campbell-Hausdorff formula, cf. Chapter III a)
$+_L$	addition on a uniform subgroup, cf. Chapter III b)
$[\ ,\ ]_L$	Lie bracket on a uniform subgroup, cf. Chapter III b)
$L(H)$	$\mathbb{Z}_p$-Lie algebra assigned to H, cf. Chapter III b)
$\mathcal{L}(P)$	Lie algebra assigned to P, cf. Chapter III b)
$\Lambda(H)$	$\log(H)$, cf. Chapter III c)
$d(N)$	number of generators of N, cf. Chapter IV
$\mho_i(N)$	$\mho_1(N) = H^p$, $\mho_{i+1} = \mho_1(\mho_i(N))$, cf. Chapter IV
$\Omega(N)$	$= \langle n \in N \mid n^p = 1 \rangle$

Φ	root system, cf. Chapter V
Q	weight lattice, cf. Chapter V
$G(\Phi, K)$	Chevalley group, cf. Chapter V
$(A, ^\circ)$	algebra A with involution $^\circ$, cf. VI a)
A^-	$= \{x \in A \mid a^\circ = -a\}$ for some order or algebra A with involution $^\circ$, cf. VI a)
$C_{PL}, c_{PL}, C_{LP}, c_{LP}$	Cayley maps, cf. Chapter VI a)
$\mathrm{rad}(\Lambda)$	radical of some order Λ
Ω	maximal order of a division algebra D, cf. Chapter VI a)
$\mathbf{P}$	$\mathrm{rad}\Omega$, cf. Chaper VI a)
$d_i(a_1, \ldots, a_n)$	matrix defined in (VI.10)
$J_i^- = J_i^-(\Lambda)$	$= (\mathrm{rad}\Lambda)^i \cap A^-$ (VI.4)
$U_i(\Lambda)$	$(1 + (\mathrm{rad}\Lambda)^i) \cap U$ (VI.4)
B_r	r^{th} dimension matrix, cf. (VI.15)
F	Gram matrix cf. after (VI.16)
$K_{ab,p}$	maximal abelian extension of exponent p of a local field K, cf. Chapter IX
$\mathcal{K}_i(K)$	division algebra of dimension i^2 over the center K cf. XI a), e)
$u_i, m_i, \widetilde{m_i}$	cf. Chapter XII
S_q	Nottingham group, cf. XIII a)

XVII Index